Computation Platforms for Multi-access Edge Computing

Hideharu Amano
Editor

Computation Platforms for Multi-access Edge Computing

Springer

Editor
Hideharu Amano
Systems Design Laboratory
Graduate School of Engineering
Bunkyo-ku, Tokyo, Japan

ISBN 978-981-96-8934-7 ISBN 978-981-96-8935-4 (eBook)
https://doi.org/10.1007/978-981-96-8935-4

This Springer imprint is published by the registered company Springer Nature Singapore Pte Ltd.
The registered company address is: 152 Beach Road, #21-01/04 Gateway East, Singapore 189721, Singapore

Foreword

It is my great pleasure to write the preface to this book, written by Hideharu Amano, one of Japan's leading computer architects and one of the research directors of the JST CREST project, which I supervised.

The purpose of CREST is promoting unique, internationally high-level basic research to address the important problems facing our country and to produce creative, conspicuous, and innovative technology seeds (new technology seeds) based on new scientific knowledge that contributes to scientific and technological innovation that can transform the society and economy. Professor Amano was the team leader of the CREST "Development of an integrated multi-node system for multi-access edge computing."

The research team proposed innovative cloud-edge information processing and designed and implemented a Multi-access Edge Computing (MEC) system. They developed hardware, OS, and applications themselves, and verified its effectiveness in several significant application fields. Therefore, the team received very high reputation not only in Japan but also in the world-wide research community.

This book shows the fruitful achievements of their research. It is definitely my honor to recommend it to all the researchers and engineers who are interested in the near future IT systems and their applications.

D.E. Vice President, The University of Tokyo — Shuichi Sakai

Preface

Our society is moving toward a super-smart society called Society 5.0, where advanced information technology will be deeply integrated into everyday life. In Society 5.0, we will deal with huge amounts of different types of data. To handle this, we need not only better traditional computing methods but also cutting-edge technologies like artificial intelligence (especially deep learning), quantum computing, and optical computing.

Real-time technology will also become more important, as it allows us to quickly solve real-world problems and respond to changing social needs. In addition, we need new technologies for hardware, circuits, system architecture, and software to make these advanced systems work efficiently, use less energy, and be secure and reliable.

To support this, the JST CREST program called "Technology for Computing Revolution for Society 5.0" was started in 2018 led by Prof. Shuichi Sakai. Its goal is to develop completely new types of computing technology, not just improvements on existing ones, to help build this future society. Eight research teams have been selected and are being actively pursued.

This book compiles the research achievements of the "Development Team for a Multi-node Integrated System for MEC." MEC (Multi-access Edge Computing) aims to deploy computing resources at 5G base stations to offload and anonymize computations performed on edge devices. By doing so, it avoids the computational load and network congestion associated with cloud-based processing, as well as the risks of transferring private information. Leveraging the powerful transmission capabilities and low latency of 5G, MEC enables near real-time processing, making it effective for controlling robots and drones, managing factory operations, implementing smart cities, and supporting autonomous driving.

A distinguishing feature of this team is its vertically integrated architecture, covering everything from applications to devices. Their scope extends from real-world implementations such as smart cities and robot control in nursing facilities to middleware for MEC, multi-FPGA systems as computing resources, high-level synthesis CAD tools for system design, and dedicated ASIC chips. This book includes the achievements of groups other than the high-level synthesis group. The Wakabayashi Group, which developed the high-level synthesis CAD, is currently integrating their tool into NEC's commercial high-level synthesis tool, CyberWorkBench. Due to the large amount of unpublished material involved, their work has been excluded from this book.

The structure of this book is as follows:

- Chapter 1 is for the introduction of MEC and the goal of this project.
- Chapter 2 describes the multi-FPGA system that was developed for this project with various techniques to build FPGA clusters.
- Chapter 3 discusses the middleware for MEC and applications such as social implementation in nursing care facilities.
- Chapter 4 focuses on the roles of MEC, such as anonymization, data collection, and selection, and describes the technologies for utilizing MEC in the construction of smart cities.
- Chapter 5 discusses the realization of Robot Audition, one of the most promising applications expected to run on MEC. It introduces an attempt to implement HARK, a widely used platform, on MEC.
- Chapter 6 presents a new FPGA device, SLMLET, which is more cost-effective and power-efficient than conventional FPGAs and features communication links for multi-node systems. It is an ambitious chip equipped with variable-structure IP, differing from traditional FPGAs.

During the course of the project, technologies similar to MEC, such as AI-RAN by SoftBank, have reached a practical level through platform providers. The FPGA board developed in this project has been commercialized, and the FPGA IP developed for the new chip is also being considered for commercial use. The findings related to smart cities have been reflected in standard specifications, and the high-level synthesis CAD has been integrated into a commercial tool. Although the project has concluded, its outcomes continue to contribute in various ways to the realization of a Society 5.0 society. This book places greater emphasis on providing an overview rather than delving into detailed technical explanations. For technical details, please refer to the many papers listed in the references.

I sincerely hope that this book will be helpful to the readers in their research and development efforts.

Tokyo, Japan
April 2025

Hideharu Amano

Acknowledgments The authors truly appreciate Akifumi Sakamoto of JST for his great effort in managing our project. This project has been sustained by the efforts of numerous researchers and students. I would like to take this opportunity to express my heartfelt appreciation.

Chapter 2
Researchers: Tomohiro Kudoh of AIST, Michihiro Koibuchi of NII, Ryousei Takano of AIST, Akram Ben Arhmed of AIST, Wei Kaijie of Keio Univ., Hiroki Matsutani of Keio Univ., Masaaki Kondo of Keio Univ./RIKEN, Ryuta Kawano, Kazuei Hironaka of Hitachi, Hayate Okuhara of NSU, Kazuhiro Wakabayashi of U-Tokyo, Wataru Takahashi of NEC, Ryota Yasudo of Kyoto Univ., and Masashi Aono of Amoeba Energy
Students (Keio Univ.): Kensuke Iizuka, Kohei Ito, Tomoki Shimizu, Ayaka Ohwada, Kazusa Musha, Keita Azegami, Yugo Yamauchi, Hideto Kayashima, Takeharu Ikezoe, Aika Kameyama, Hiroaki Suzuki, Koki Honda, Haruna Takagi, Chen Yu Cheng, Aoi Hiruma, Huan Peng Yu, Pengyu Huang Zhongyang Hou, Ziquan Qin, Ullah IMDAD, Yuko Yamakura, Takumi Inage, Souhei Takagi, and Yasuyu Fukushima

Chapter 3
Collaborators: Youichi Ishiwata, Kaoru Suzuki, Harumi Watanabe, Eisuke Okazaki, Gai Nagahashi, Yuta Hasuhara, Hayato Mori, Daniel Pinheiro Leal, Ludi Wang, Masato Fukui, Koshirou Kubota, Rika Honda, Koki Saito, Hajime Sasaki, Koki Nagahama, Koki Higashi, Kouki Ozawa, and Natsuho Sannomiya

Chapter 4
Students: Yuiko Sakuma, Saki Korosue, Shogo Shimahara, Moena Honda, Koutaro Yamamoto, Hiroki Hiraga, Tokimasa Toyohara, Yohei Nanba, Akari Sato, Keiten Han, Kentaro Morise, Yuto Chikaoka, Hiroshi Nakano, and Takuma Fukui

Chapter 5
Researchers: Masayuki Takigahira (Honda Research Institute Japan), Naoya Terakado (Honda Research Institute Japan), Katsutoshi Itoyama (Tokyo Institute of Technology/Honda Research Institute Japan, currently Institute of Science Tokyo), Yui Sudo (Honda Research Institute Japan, currently SB Intuitions), Kenji Nishida (Tokyo Institute of Technology, currently Kaishi-Sozo University) Monikka Roslianna Busto (NTT), Akiko Masaki (NTT), and Takeharu Eda (NTT)
Students: Yanke Long (Institute of Science Tokyo, currently Advanced Media), Muhammad Shakeel (Tokyo Institute of Technology, currently Honda Research Institute Japan), Atsuo Hiroe (Institute of Science Tokyo), Ragib Amin Nihal (Institute of Science Tokyo), Aoto Yasue (Tokyo Institute of Technology, currently Nomura Research Institute), and Tsubasa Yano (Institute of Science Tokyo)

Chapter 6
Researchers: Motoki Amagasaki of Kumamoto Univ., Qian Zhao of Kumamoto Univ. (currently ByteDance), Takuya Kojima of Tsukuba Univ., Hayate Okuhara of NSU
Students (Keio Univ.): Yanai Yosuke
Students (Kumamoto Univ.): Kyosei Yanagida, Ryo Araki, Hiroaki Koga, Yuya Nakazato, Masahiro Sumita, Izumi Kiuchi, Zhan Yan, Koudai Takeno, and Hayato Abe

This work was supported by JST CREST JPMJCR19K1.
This work was also supported by VDEC in collaboration with Cadence Design Systems, Synopsys Corporation, and Mentor Graphics Corporation.

Contents

Acronyms

JST	Japan Science and Technology Agency (JST)
MEC	Multi-access Edge Computing
FPGA	Field Programmable Gate Array
CPU	Central Processing Unit
GPU	Graphic Processing Unit
ASIC	Application Specific Integrated Circuits
AI-RAN	Artificial Intelligence Radio Access Network
NEDO	New Energy and Industrial Technology Development Organization
AXI	Advanced Extensible Interface
SoC	System-on-a-Chip

Chapter 1
Introduction

Hideharu Amano

1 Cloud and Edge Computing

In today's IT society, computing can be broadly categorized into Edge Computing, which is performed close to the user, and Cloud Computing, which is carried out in large-scale data centers. Edge Computing is performed on computing devices equipped in user-owned mobile devices, robots, drones, vehicles, IoT (Internet of Things) devices such as sensors, and surveillance cameras. PCs and high-spec gaming machines are sometimes considered Edge devices as well, though they possess significant computing power. However, much of Edge Computing is executed on less powerful devices with significant limitations due to power budgets. On the other hand, Cloud Computing is performed on powerful computing devices located in large-scale data centers.

Data centers can perform massive computations at low cost by sharing power, air cooling, storage, and networks across numerous computing devices. Big data processing, which involves handling vast amounts of data, can only be conducted in data centers where such data is aggregated. Additionally, the training phase of AI systems using large-scale language models (LLMs), which have been rapidly gaining popularity, relies on powerful GPUs installed in data centers. In this way, the Cloud and Edge collaborate to divide and execute the computational tasks required by modern society.

If we define a hunting society as Society 1.0, an agricultural society as Society 2.0, an industrial society as Society 3.0, and an information society as Society 4.0, we can say that we are currently living in Society 4.0. At the Japan Science and Technology Agency (JST), Society 5.0 is defined as a human-centered society that

H. Amano (✉)
Systems Design Laboratory, Graduate School of Engineering, Bunkyo-ku, Tokyo, Japan
e-mail: hunga@dlab.t.u-tokyo.ac.jp

H. Amano (ed.), *Computation Platforms for Multi-access Edge Computing*,
https://doi.org/10.1007/978-981-96-8935-4_1

achieves both economic development and the resolution of social challenges through a highly integrated system of cyber and physical spaces. To realize this vision, JST launched Innovative Computing Technologies Supporting Society 5.0 in 2020.

The computing systems required to support the Society 5.0 era cannot rely solely on the current Cloud-Edge architecture. For many IoT devices, there is a significant gap between the computational power available in the Cloud and that at the Edge. As a result, most calculations are sent from Edge devices to the Cloud, where they are processed. As a large volume of data is sent to the Cloud, the network becomes congested, latency can sometimes increase significantly, and the situation becomes unpredictable. Consequently, it is impossible to perform timing-critical control tasks, such as those for robot control, drone operations, or autonomous driving. Furthermore, the data collected by Edge devices often includes information related to privacy, making it unsafe to send it to the Cloud in its raw form. To address these issues, the concept of Between-Cloud-and-Edge Computing, which places intermediate computational resources between the Cloud and the Edge, has been under consideration for a long time.

2 Between-Cloud-and-Edge Computing

Fog Computing, proposed by Cisco [1], involves placing computational resources at network relay points. Since this concept was introduced before the widespread development of IoT devices using wireless technologies like Wi-Fi, the computational resources were placed in network routers and similar devices. As an alternative approach that places computational resources closer to the edge, Mist Computing [2] was proposed, but this concept did not gain widespread adoption.

With the advancement of 5G wireless technology, a wide range of devices can now connect with high bandwidth and low latency. MEC (Multi-access Edge Computing) [3, 4] is an initiative to enable timing-critical control by placing computational resources at base stations. Figure 1.1 illustrates the differences between traditional computational infrastructures and MEC. MEC (Multi-access Edge Computing) is designed for control within a limited range, targeting applications such as carrier control within factories and robot or drone control within specific premises. It also handles various tasks related to control, including image and voice recognition processing. Additionally, MEC can be used for anonymization processes, security measures, and data transfer protocol conversion before sending data to the Cloud.

The concept of MEC is sometimes referred to as Mobile Edge Computing, with an expanded scope aimed at applications like autonomous driving and traffic control over broader areas.

In 2023, accelerated by the rapidly advanced AI techniques, SoftBank has actively applied this approach in its AI-RAN (Radio Access Network), which envisions high-performance wireless networks. It integrates powerful GPU-level computational capabilities at base stations, sharing these resources between essen-

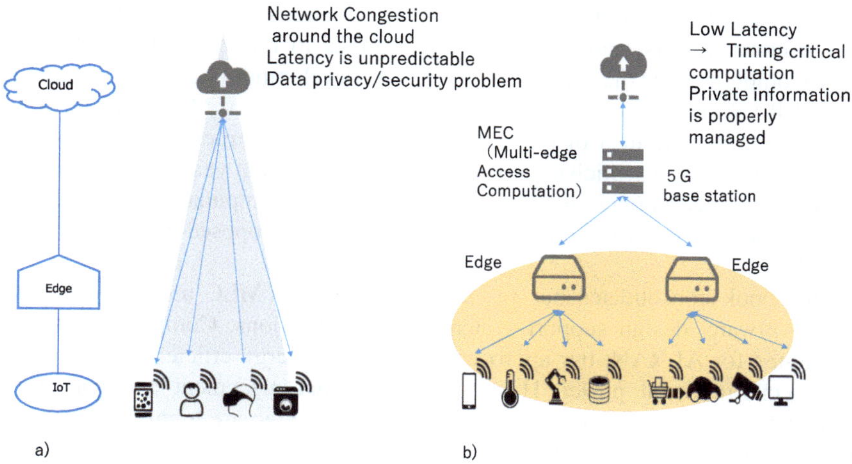

Fig. 1.1 The current cloud-edge computing and between-cloud-and-edge computing. (**a**) Traditional computing with edge/cloud. (**b**) Between-edge-and-cloud computing

tial communication service processing for base stations and rapidly advancing AI processing.

AI-RAN leverages ultralow latency for traffic services such as autonomous driving using multimodal AI for traffic understanding, ensures security through high isolation, and features capabilities like 8K video processing using high data transfer capacity and video processing with numerous cameras [5].

As of 2025, an experimental environment has been established at Keio University's Shonan Fujisawa Campus. Utilizing a platform called AITRAS, equipped with NVIDIA's Grace-Hopper GH200, experiments such as security services performed by robots are being conducted.

3 Development Project for a Multi-node System for MEC

These Between-Cloud-and-Edge Computing initiatives place the computational infrastructure at relay points to achieve common goals: real-time control utilizing low latency, enhanced security leveraging high isolation, and large-volume transfers unaffected by congestion in the cloud. Over time, the location of these relay points has shifted from WAN relay stations to 5G wireless base stations, and the computational infrastructure has evolved from simple servers to various computing devices.

Our "Development Project for a Multi-Node System for MEC" began in 2019 as part of JST's "Innovative Computing Technology to Support Society 5.0" initiative [6]. This five-and-a-half-year project aims to explore the optimal computational infrastructure for deployment at wireless base stations, develop software, create design environments, and implement applications to realize MEC (Multi-access Edge Computing). At the start of the project, AI development was still in its infancy, and AI-RAN did not yet exist. Consequently, the concept of MEC was not widely recognized, and the idea of placing computational infrastructure at base stations itself was met with skepticism. However, by 2025, the final year of the project, AI-RAN, which incorporates significantly more powerful computational infrastructure at base stations than we initially envisioned, has been proposed by major platform providers.

This book consolidates our research outcomes on MEC architectures, hardware, software, design support systems, and applications. Compared to what is envisioned for AI-RAN, our approach emphasizes energy efficiency and employs lower computational power. However, the optimal placement and capacity of computational resources between the cloud and the edge are highly dependent on network configurations and the nature of target applications. We believe that our research has played a pioneering role in Between-Cloud-and-Edge Computing and will significantly contribute to its future development.

The structure of this book is as follows: Chap. 2 shows multi-FPGA platforms as a host of MEC architecture. Various techniques and examples of implementation on multi-FPGA platforms are introduced. Chapter 3 focuses on middleware for MEC. MEC-RM (MEC Resource Manager), a general framework for MEC, is explained. Chapter 4 is for the application for a smart city and practical examples of smart community platforms. Chapter 5 describes the automatic speech recognition with HARK system implemented on the MEC architecture. Finally, Chap. 6 shows an example of a real ASIC implementation with original embedded FPGA IPs.

References

1. F. Bonomi et al., Fog computing and its role in the Internet of Things, in *Proceedings of the First Edition of the MCC Workshop on Mobile Cloud Computing*. MCC '12 (Association for Computing Machinery, Helsinki, 2012), pp. 13–16. ISBN: 9781450315197. https://doi.org/10.1145/2342509.2342513
2. P. Galambos, Cloud, fog, and mist computing: advanced robot applications. IEEE Systems Man Cybern. Mag. **6**(1), 41–45 (2020). https://doi.org/10.1109/MSMC.2018.2881233
3. Y.C. Hu, M. Patel, D. Sabella, N. Sprecher, V. Young, Mobile edge computing: a key technology towards 5g. ETSI White Paper No.11. 2015
4. ETSI, Multi-access edge computing (2018). Accessed 30 Jun 2024. https://www.etsi.org/images/files/%5C%5CETSITechnologyLeaflets/MultiAccessEdgeComputing.pdf
5. AI-RAN alliance, Redefine what networks can transform (2025). Accessed 30 Jan 2025. https://ai-ran.org/
6. JST CREST, Technology for Computing Revolution for Society 5.0 (2019). Accessed 30 Jan 2025. https://www.jst.go.jp/kisoken/crest/en/research%5C_area/ongoing/areah30-4.html

Chapter 2
Multi-FPGA Platforms as a Host of MEC Architecture

Hideharu Amano, Takuya Kojima, and Mitaro Namiki

1 Why FPGAs Are the Best Candidate for MEC?

FPGA (Field-Programmable Gate Array) is a reconfigurable LSI traditionally used in digital circuits focused on hardware processing, such as prototyping for ASIC design and network interfaces. However, with the increasing demand for AI processing, FPGAs have also been utilized as accelerators connected to CPUs to enhance certain functions.

In 2020, Intel acquired Altera, and in 2023, AMD merged with Xilinx, with both companies initiating efforts to use FPGAs as AI accelerators. However, as GPU performance advanced rapidly, FPGAs struggled to compete in terms of performance relative to cost. Consequently, they could not take the lead as AI accelerators, and in 2024, Intel decided to spin off its FPGA business.

However, as a computational element equipped at base stations for MEC (Multi-access Edge Computing) purposes rather than in the cloud, FPGAs offer the following advantages:

- They have higher power efficiency than GPUs, making them well suited for base stations where the power supply is constrained.

H. Amano (✉)
Systems Design Laboratory, Graduate School of Engineering, Bunkyo-ku, Tokyo, Japan
e-mail: hunga@dlab.t.u-tokyo.ac.jp

T. Kojima
Department of Computer Science, Institute of Systems and Information Engineering, The University of Tsukuba, Tsukuba, Japan

M. Namiki
Department of Electrical Engineering and Computer Science, Tokyo University of Agriculture and Technology, Tokyo, Japan

H. Amano (ed.), *Computation Platforms for Multi-access Edge Computing*,
https://doi.org/10.1007/978-981-96-8935-4_2

- Since computations are performed in hardware, processing times exhibit less variability, making them ideal for real-time processing with strict timing requirements.
- Base stations are inherently equipped with FPGAs for wireless communication, which can be extended for computational use.
- FPGAs can integrate switching and interface functions, allowing for easy connection to other devices or additional FPGAs, enabling scalable expansion.

The main issue with FPGAs is that their hardware resources are limited, which can restrict the range of applications they can support. However, in the case of MEC (Multi-access Edge Computing), the applications can be somewhat tailored to the system. Connecting multiple FPGA boards to construct a multi-FPGA system makes it possible to expand the amount of hardware resources as needed.

Thus, a scalable multi-FPGA system or FPGA cluster could serve as a viable solution for providing computational resources for MEC.

2 The M-KUBOS Board

As a computational element for MEC, we developed an FPGA board called M-KUBOS cooperated with PALTEK. The overall architecture of M-KUBOS is based on a prototype cluster called FiC developed in an NEDO project [1].

Its photo and the overview are shown in Fig. 2.1. For operating stand-alone at the base station, the SoC (System-on-Chip) FPGA needs to be provided. In the M-KUBOS board, Xilinx's SoC-FPGA Zynq Ultrascale+ xczu19eg, two sets of 4 GB DRAM, and various high-speed serial and other interfaces are equipped. Zynq xczu19eg is the largest scale model in this series which provides hardware resources shown in Table 2.1. Its PS (Processing System) section includes an ARM Cortex-A53 Quad-Core and Cortex-R5 Quad-Core. The PL (Programmable Logic) section provides a logic capacity comparable to large-scale FPGAs among mid-scale devices, making it suitable for use as an accelerator.

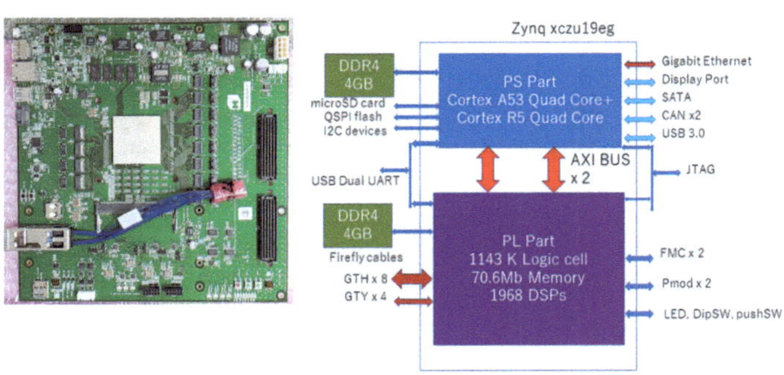

Fig. 2.1 M-KUBOS board

Table 2.1 The resources of PL part in M-KUBOS board

DSP	BRAM	LUT	FF	URAM	Total system logic cells (K)
1968	34.6 Mb	523000	1045000	36 Mb	1143

Table 2.2 Detailed I/O interfaces of M-KUBOS

Item	Specification
Memory	PS: 4 GB DDR4-2400
	PL: 4 GB DDR4-2400 SODIMM × 1
Serial Link	4x GTY 4TX/4RX (max 28.125 Gbps)
	4x GTH 8TX (max 16.3 Gbps)
	4x GTH 8RX (max 16.3 Gbps)
I/O	USB3.0 × 1
	USB-UART × 1
	1 Gb Ethernet (RJ45)
	DisplayPort 1.2

The PS and PL sections are connected via two sets of buses, enabling data exchange through DMA transfer using AXIS and shared memory using AXI4lite. The DRAM is directly connected to both the PS and PL sections, allowing access with a 32-bit width for the PS and a 64-bit width for the PL. The serial interfaces connected to the PL consist of eight sets of GTH and four sets of GTY. The detailed specifications of the board are shown in Table 2.2.

3 The M-KUBOS Cluster

The M-KUBOS board can be configured into a multi-FPGA system by connecting high-speed serial links. The M-KUBOS boards are interconnected using Samtec's FireFly™ cables [2]. Each cable connects eight GTH links in a unidirectional configuration. Therefore, two FireFly cables are required for bidirectional connections. Thus, we can form eight four-lane bidirectional links and various types of topologies by using them. The simplest example of a cluster structure of M-KUBOS is shown in Fig. 2.2. Six boards form a bidirectional ring topology.

Although the topology is simple, it is well suited to the streaming processing that frequently appears for MEC.

3.1 Shell–Role (Crust–Core) Architecture

Since an FPGA is a reconfigurable device, we can push every hardware module to form a cluster structure. However, to behave as a computation element in MEC, it must also provide a user-defined application part.

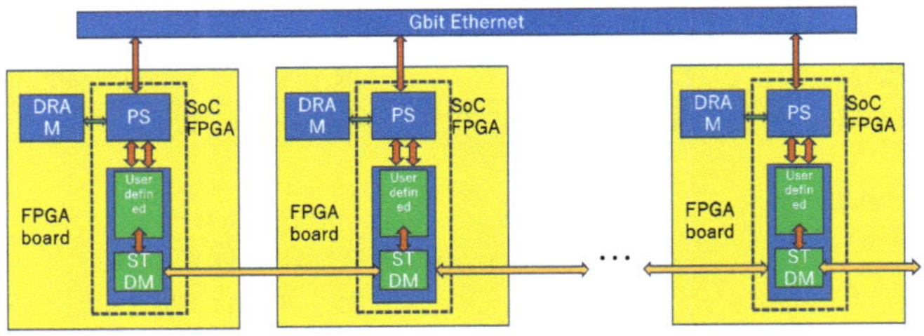

Fig. 2.2 The simplest example of M-KUBOS cluster

In other words, each M-KUBOS board should ideally be designed with two distinct parts: a fixed section with predefined functionalities, such as the DRAM interface, data transfer switches for communication with other boards, and interfaces with the PS unit, and a flexible section where applications developed by programmers can be deployed. Here, we will refer to the former as the Shell(Crust) section and the latter as the Role(Core) section.

While multiple types of Shell sections may need to be prepared based on the nature of the applications in the Role section, the design of these Shell sections themselves should remain unchanged. Programmers can write their applications using HLS (High-Level Synthesis) and deploy them in the Role section to execute. This eliminates the need for programmers to design interfaces, allowing them to focus entirely on the design of individual applications.

This design approach is referred to as the Shell–Role architecture. A similar design approach has long been adopted in ASIC design, where it is known as structured ASIC [3]. Unlike structured ASIC, we can select various types of Shell structures so as to keep the area for the user-defined "Role" as large as possible. We prepared the following shells:

- S (Single): Only interfaces between the PS including the DMA controller and control registers are provided.
- SwD (Single with DRAM): DRAM controller (MIG) is added to S.
- M (Multiple): Network interface IPs are provided to connect with the other M-KUBOS board.
- F (Fully equipped): All IPs, network interface, MIG, and DMA controller are equipped.

Before explaining the details of each shell, the structure of the switch and networks is introduced.

3.2 STDM (Static Time Division Multiplexing) Switch

FPGA clusters are primarily proposed for use in cloud environments, with most relying on packet-switched networks. This approach offers high flexibility and ensures efficient line utilization regardless of congestion levels. However, the M-

KUBOS cluster adopts an STDM (Static Time Division Multiplexing) approach. In this method, communication lines are divided into fixed time slots, allocating specific time intervals to data transmission [4, 5].

Also in this method, communication channels are statically assigned to time slots, creating a pseudo-dedicated line between communicating modules. This approach is well suited for timing-critical processes, as it guarantees bandwidth and latency between modules. Additionally, it requires smaller buffer sizes for implementation, making it ideal for MEC clusters.

However, there are drawbacks: If many lines are required and the data exchange per line is minimal, an increase in the number of time slots can reduce the bandwidth per line and increase latency. In the M-KUBOS cluster, each GTH serial communication channel is used as an individual lane, which helps reduce the number of time slots needed for each lane.

As introduced before, each GTH line has 8 bundles (32 channels) connected to other boards. Each serial channel is connected to Xilinx serial transfer IP Aurora, and 64b/66b encoded data with ECC are transformed into 85 bits for each 100-MHz clock cycle. (Actually, data of 170 bit width are transferred every two clocks). That is, the effective bandwidth of the FiC-SW is 8.5 Gbps for each channel. The 85-bit data are exchanged in the STDM switch according to the slot number and the routing table, transformed into the serial data again, and then transferred to the next board through the Aurora IP. Here, we focus on the structure of the STDM switch. We used a relatively low clock frequency of 100 MHz in the STDM switch to avoid the trouble. The system operates with up to nine inputs and nine outputs, the maximum size achievable on the board, simply by modifying the parameters within the same design.

Figure 2.3 shows an example of switches provided in the M-KUBOS cluster. The shaded part shows the Shell part, while the white area represents the user-defined role part. Two parts can work at different clock frequencies and are connected with the AXI stream interface like other IPs.

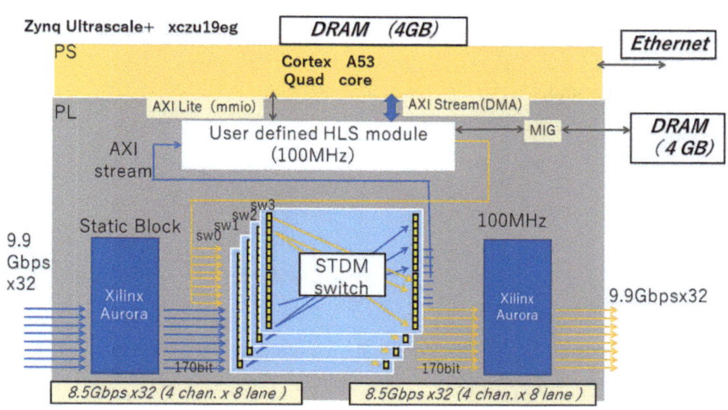

Fig. 2.3 Switches in the Shell part

3.3 Performance and Resource Usage

The bandwidth of a serial link provided for the M-KUBOS cluster is 8.5 Gbps = 1.065 GB/sec when a slot of the STDM is used. When S slots are used, it is divided by S. Since the M-KUBOS cluster provides eight links between connected boards, the bandwidth can be increased by using Link Aggregation and Slot Distribution. The maximum latency between boards is represented with

$$2S(L-1) + (60 + 2S)H \quad clock\ cycles,$$

where S is the slot number, L is the flits to be transferred (170 bits/1-flit), and H is the hop number. The STDM switch and serial links of M-KUBOS are compatible with those of FiC, and the performance is analyzed in detail [1, 4]. The performance depends on the number of slots. Thus, the optimization algorithm proposed in [6] is used.

First of all, the transfer latency between PS and PL is evaluated and shown in Fig. 2.4. The performance depends on the language that controls the data transfer and whether DMA is used. If DMA is used and a large block of data is transferred, the transfer latency is independent of the language. However, if the data size is small and the DMA is not used, the difference between C and Python becomes large.

Figure 2.5 shows the throughput from the PS part of a board to the PS part of the next board (next board). The throughput is saturated when about 60 KB data chunk is transferred. When the number of intermediate boards is increased, the latency is increased according to the expression shown above.

Figure 2.6 shows the resource utilization of each Shell. Although F requires the most resources, LUTs, LUTRAMs, and FFs are approximately 10% or less. BRAM used for buffers between boards consumes approximately 20% for M and 25% for F, and it decreases the resources for the user-designed part. We must reduce the unnecessary buffer used in the network IP or distribute it to the URAM that is not used in the Shell part.

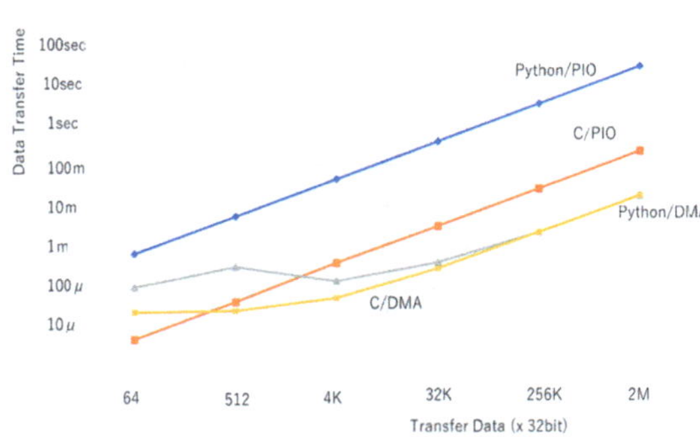

Fig. 2.4 Data transfer performance between PS and PL

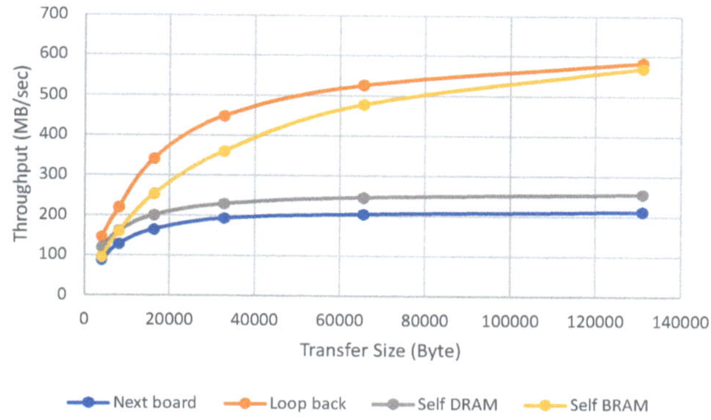

Fig. 2.5 Data transfer performance between PS and PL

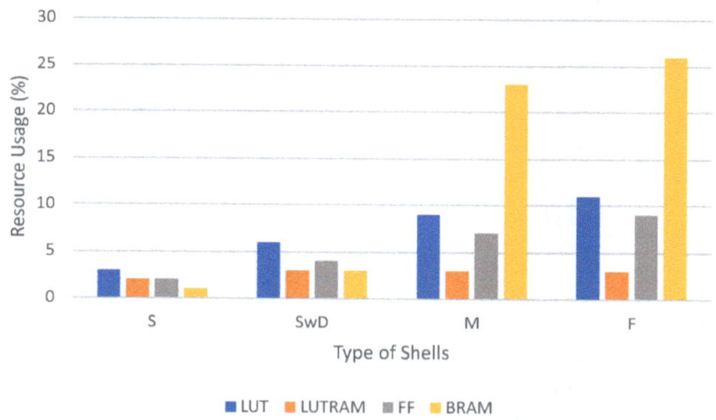

Fig. 2.6 Resource usage of each Shell

3.4 PYNQ as a Software Platform

Computing nodes for MEC run common Linux as a server used by multiple tenants. To use the PS of Zynq as a server, we introduced PYNQ to the M-KUBOS board. PYNQ (Python productivity for Zynq) [7] is an open-source software platform developed by Xilinx mainly for Zynq boards. As shown in Fig. 2.7, PYNQ employees commonly used Ubuntu as an OS kernel. It includes device drivers to enable configuration as a form of "overlays." Additionally, Xilinx's standard IPs (intellectual properties) and user IPs in the PL can be accessed from the PS through the MMIO (memory-mapped IO) by using the APIs. The PYNQ program is a combination of software PYNQ libraries in the PS and hardware IPs in the PL through the MMIO and DMA. Hardware in the PL is designed by connecting modules with Vivado block design. If the target application can be formed by connecting Xilinx's standard IPs and predesigned user IPs, the programmer's job is just connecting and accessing them with the aid of the Python and Jupyter Notebook

Fig. 2.7 PYNQ software stack

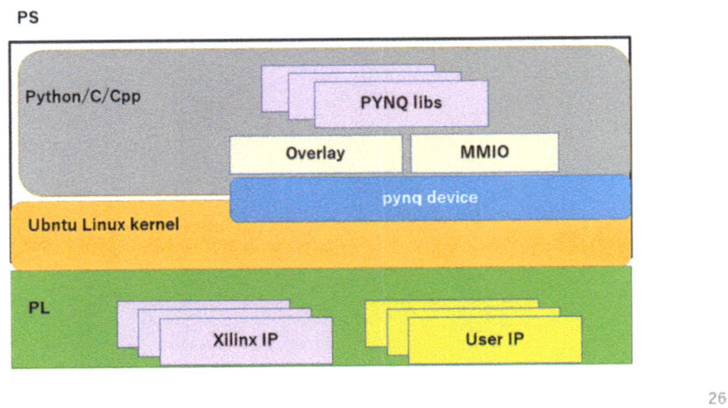

Table 2.3 Examples of available RESTful APIs

API	Method	Function
FPGA	POST	Configure the PL
	GET	Get the configuration information
switch	POST	Set the routing table of STDM switches
status	GET	Get the status of M-KUBOS board
regread	GET	Read the data of the specified address
regwrite	POST	Write the data into the specified address
runcmd	POST	Execute the specified command for M-KUBOS

software environment. Of course, the hardware designers must design their own accelerators by using Vivado HLS or HDL and register them into the IP catalog. Although various skills are required to use PYNQ, it is popularly used for developing AI applications on the PYNQ-Z1 board. From the viewpoint of MEC, PYNQ is advantageous since it can be used as a software platform for a multi-tenant server.

To connect multiple boards and manage them, RESTful APIs and their local manager mkuboswww are implemented on each board. Each RESTful API shown in Table 2.3 is represented with a URL and called with a simple http request. mkuboswww is written in Python and uses Flask as an application framework. Some commands including overlays of the configuration need the superusers access permission, but *mkuboswww* cannot directly execute such requests. To execute such commands, the RPC server is invoked with superuser rights and executed indirectly. This allows users to manage the M-KUBOS cluster remotely without obtaining superuser permission. Designers can describe their requests including configurations, STDM table settings for routing, and starting of the user programs on each node in the JSON script. On the remote server, mkubosmgr reads the script and distributes the requests to the specified board.

The relationship between *mkuboswww*, *mkubosmgr*, and basic PYNQ layers is shown in Fig. 2.8. Since PYNQ is formed with commonly used Ubuntu, *mkuboswww* and RESTful APIs can be implemented as common servers.

In the M-KUBOS cluster, a job scheduling system is implemented to enable shared use by multiple users. The system uses the Slurm Workload Manager [8],

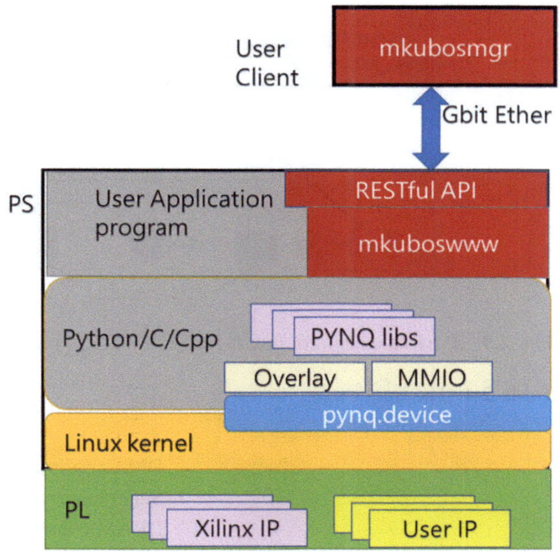

Fig. 2.8 Software for PYNQ cluster management

an open-source job scheduling system with a proven track record in HPC clusters, deployed on the Management Server shown in Fig. 2.2.

Additionally, by utilizing Slurm's node partitioning feature, predefined node groups (partitions) consisting of interconnected M-KUBOS nodes are established based on the number of nodes in the M-KUBOS cluster. This setup enables the allocation of the required number of M-KUBOS nodes according to the demands of each job execution. More advanced management systems are described in Chap. 3.

3.5 The IP-Based Design

The design concept of PYNQ is a combination of software/hardware IPs, and the design on the PYNQ cluster is just its extension; that is, we treat the STDM switches/Aurora IPs as network IPs. The designer's job is to develop IPs by using Vivado HLS or HDL and connect with other IPs. If the IPs needed to communicate are located on different boards, the network IPs are connected with the Vivado block design editor. The concept of the IP-based design is shown in Fig. 2.9. The ARM CPU of the PS is also treated as an IP like other Zynq designs.

Table 2.4 shows IPs commonly used for designs of the M-KUBOS cluster. They are supported by Xilinx, for a single board of M-KUBOS and for building a cluster. Like other Zynq boards with PYNQ, AI platforms, FINN [9], and VTA are also supported but omitted from the table. Note that the network interface IP includes an STDM switch, SERDES, and Aurora interface concealed from the designer. Most

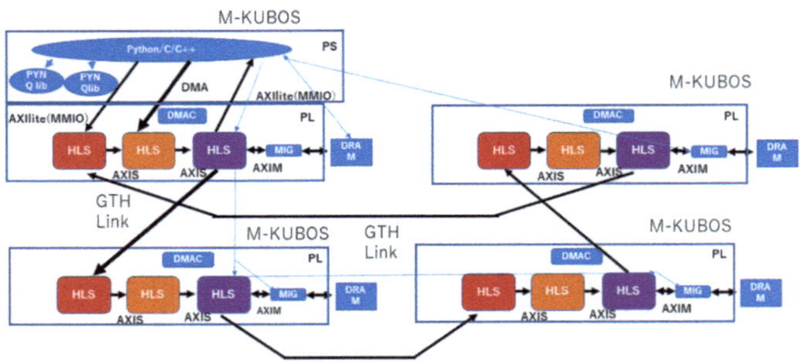

Fig. 2.9 IP-based design for the M-KUBOS/PYNQ cluster

Table 2.4 The frequently used IPs for the design

IP	Design	Supported
Zynq	Embedded	Xilinx
DMA controller	Embedded	Xilinx
Smart connect	Embedded	Xilinx
DRAM controller (MIG)	Embedded	Xilinx
Control registers	HDL	M-KUBOS
BRAM/interface	HDL	M-KUBOS
Timer	HDL	M-KUBOS
PMOD/LED interface	HDL	M-KUBOS
Network interface	HDL	Cluster
Message sender	HLS	Cluster
Message receiver	HLS	Cluster
Merger	HDL	Cluster
Splitter	HLS	Cluster

IPs are connected with the AIX4 stream interface except the connection with Zynq and DDR SDRAM.

3.6 *Implemented Application and Benchmarks*

On M-KUBOS and its prototype FiC, various types of applications except those treated in the later sections: Recurrent Neural Network [10], LeNet [11], ResNet [12], Vision Transformer [13], Genome Matching [14], and Amoeba-SAT solver [15]. We picked up the following application programs for evaluating the M-KUBOS cluster.

(a) ResNet-50: Image recognition is often required for controlling robots and drones, and one anticipated application of the MEC cluster is image recognition through deep learning. ResNet [12] is a representative convolutional neural network characterized by the extensive use of bypass layers.

In this study, we implemented ResNet-50, which processes 224 × 224 pixel images, by distributing its layers across four nodes. The implementation involves inputting images into the PL of board 0 via the PS, then processing the data sequentially in a pipeline manner through the M-KUBOS serial links as follows: board 0 → board 1 → board 2 → board 3. Finally, the results are retrieved by the PS of board 3.

While image data is transmitted in a streaming fashion, weight data is preloaded into DRAM for use during computation. To enhance performance without compromising recognition accuracy, partial int8 quantization was applied during the implementation. For detailed information on the implementation, refer to [12].

(b) Amoeba-SAT Solver: Carrier control in factories is a representative application of MEC. Scheduling tasks in this context often take the form of constraint satisfaction problems. The Amoeba-SAT solver is a heuristic algorithm for solving constraint satisfaction problems, inspired by the behavior of the slime mold amoeba. Its application to large-scale transport scheduling problems is being explored.

The Amoeba-SAT solver evaluated here has been improved to handle a broader range of problems compared to a previously proposed method [16], though its execution speed is slower. For more details, refer to [15].

(c) Conjugate Gradient Method and FFT: The Conjugate Gradient (CG) method is commonly used as an iterative solver for large-scale matrices and was selected as a representative numerical computation involving floating-point operations. The implementation on M-KUBOS utilizes 32-bit single-precision floating-point arithmetic, targeting dense matrices (384×384) and performing computations that involve extensive broadcast communication between M-KUBOS nodes. Parallel processing was conducted using two and four nodes.

Additionally, the Fast Fourier Transform (FFT), widely used in signal processing, was included as a sample application simulating signal processing within MEC. Here, the problem involved 8192 points using 32-bit single-precision floating-point arithmetic, and parallel processing was performed on one or two nodes. This application is also characterized by a significant amount of internode communication. For further details on the implementation, refer to [5]. The specification of application programs is shown in Table 2.5

3.7 *Performance/Power Comparison*

Comparison Target: Recent PC CPUs, particularly the AMD Ryzen series, are manufactured using cutting-edge semiconductor processes, offering high performance with significantly lower power consumption compared to previous generations. These features make them a viable candidate for computing mechanisms deployed in MEC at base stations. Here, we compare the performance and system power consumption of the M-KUBOS cluster against the AMD Ryzen 7 5800X, an 8-core

Table 2.5 Benchmark programs

Appl.	Parameters	Num. of boards	Shell-type Purposes
ResNet-50	50 layers (fp32 + int8 Quantization) [12]	4	F Image recognition
Amoeba-SAT Solvers	10000 variables 43000 clauses	1	S SAT solver
CG method	384×384 fp32 Dense matrix	1, 4	M Matrix computation
FFT	1D 8192 points Single precision (fp32)	1, 2	M Image processing

Table 2.6 Execution time of benchmarks

Application	Threads/ nodes	Ryzen7 5800X	M-KUBOS cluster
ResNet-50	4	143.2 ms	13.3 ms
Amoeba-SAT	1	117207.4 ms	9339.8 ms
CG method	2	26.0 ms	11.7 ms
	4	13.7 ms	7.6 ms
FFT	1	0.094 ms	0.104 ms
	2	0.054 ms	0.061 ms

processor based on the widely available Zen 3 architecture. A comparison of the systems used for evaluation, the M-KUBOS cluster and the Ryzen 7 PC, is presented in Table 2.6.

For the ResNet-50 evaluation, we used a pretrained model employing 32-bit single-precision floating-point operations included in PyTorch 1.8.1 and Torchvision 0.9.1. For the Amoeba-SAT solver, CG method, and FFT, programs were compiled using GCC 9.3.0 with an optimization level of -O3. Parallelization for the CG method and FFT, where kernels are parallelizable, was implemented using OpenMP.

To ensure a fair comparison between CPU execution and M-KUBOS cluster execution for each application, we matched the problem sizes, parameters, the number of execution threads on Ryzen 7, and the number of execution nodes on the M-KUBOS cluster. The theoretical peak performance of 32-bit floating-point operations for the Ryzen 7 5800X, per core, is approximately 150 GFLOPS.

To compare power consumption at the system level, we measured the power usage of the Ryzen 7 and the M-KUBOS system using the power supply units of the PC and FPGA boards employed for the comparison. Current measurements were performed using a clamp meter (Model: BSIDE ACM91, accuracy: ±3%), and the power consumption during application execution was recorded.

The PC used for the comparison was configured with the minimum setup necessary to run the operating system, and power consumption was measured under these conditions.

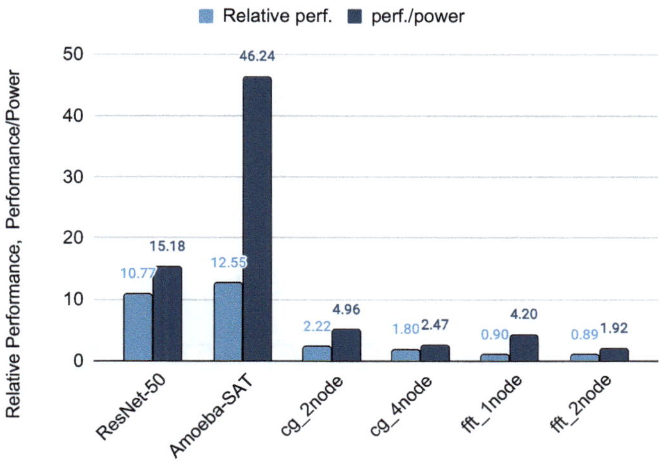

Fig. 2.10 The relative performance and power efficiency (Perf./W)

It is important to note that the Ryzen 7 used in the comparison is manufactured using a 7-nm process, which is two generations ahead of the 16-nm process used for the Zynq UltraScale+ MPSoC. As a result, this comparison involves a handicap in favor of the Ryzen 7.

Table 2.6 shows the execution times for each application, and Fig. 2.10 illustrates the relative performance and power efficiency (performance per power) of M-KUBOS compared to Ryzen. Although M-KUBOS operates at a lower frequency compared to the Ryzen 7, it achieved comparable performance for applications involving iterative numerical computations, such as CG methods and FFT. Moreover, in cases like ResNet-50 and Amoeba-SAT, it outperformed the Ryzen 7, with a maximum performance advantage of 12.56 times for Amoeba-SAT.

Power efficiency generally exceeded that of the Ryzen 7, reaching up to 46.24 times higher. Applications such as ResNet-50 and Amoeba-SAT exhibit high data parallelism and have many components of computation logic that can be parallelized as hardware, making them well suited for FPGA implementation and enabling the achievement of superior power efficiency.

4 Innovative Techniques on M-KUBOS/FiC Clusters

Link aggregation/slot distribution: In order to make use of multiple lanes for the same destination, two techniques: link aggregation and slot distribution are tried [5, 17]. The straightforward method to make the use of multiple lanes is the link aggregation. It is a method that aggregates multiple lanes and regarded as one lane [4].

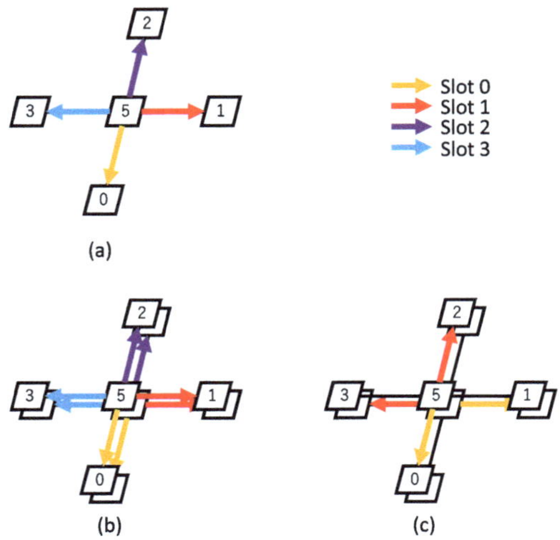

Fig. 2.11 An example of communication with one lane and applying the link aggregation and the slot distribution. (**a**) An example of communication with one lane. (**b**) The link aggregation with two lanes. (**c**) The slot distribution with two lanes

Figure 2.11a shows an example of communication when only one lane is used. In Fig. 2.11b, the central node is sending packets in four slots to neighboring boards: east, west, north, and south. In the example, packets are transferred from Node 5 to Node 0 in slot 0, to Node 1 in slot 1, to Node 2 in slot 2, and to Node 3 in slot 3.

On the other hand, Fig. 2.11b shows the case of applying the link aggregation with two lanes to Fig. 2.11a. Packets are transferred only in one lane in Fig. 2.11a, while in Fig. 2.11b packets are transferred using two lanes in parallel. Packets are transferred from Node 5 to Node 0 with both upper and lower lanes in slot 0, to Node 1 with them in slot 1, to Node 2 with them in slot 2, and to Node 3 with them in slot 3, respectively.

When the number of lanes is N, the bandwidth that can be used for each slot can increase N times by applying the link aggregation with multiple lanes. In this method, all the STDM switches on the same board have the same configuration, and the configuration of the STDM switches and the number of slots are also the same as the case of using a lane. The second method is the slot distribution. It distributes slots to multiple lanes to reduce the number of slots for a lane.

Figure 2.11c shows the case of applying the slot distribution with two lanes to Fig. 2.11a. All four slots are processed in a lane in Fig. 2.11a, while in Fig. 2.11c the number of slots is halved by distributing slots for each lane. In slot 0, packets are transferred from Node 5 to Node 0 with the upper lane and from Node 5 to Node 1 with the lower lane. In slot 1, packets are transferred from Node 5 to Node 2 with the upper lane and from Node 5 to Node 3 with the lower lane.

If the number of lanes is N, the number of slots can be reduced to $1/N$ by applying the slot distribution with multiple lanes. However, the bandwidth that can be used for packet transfer is the same as that of one lane. In this method, the STDM switches on the same board generally must have individual configurations.

When the number of slots is large and the number of packets is small, the slot distribution is sometimes more effective, while the link aggregation is superior for other most cases.

4.1 Multi-ejection STDM Switch

In circuit-switched interconnection fabrics, injection/ejection ports tend to be bottlenecks. Figure 2.12a is an example in which only a port is provided for injection/ejection. Especially, in FiC, when some packets are sent/received to/from the user logic, the same number of time slots is allocated to the link between the injection/ejection port and the user logic. For example, when the user logic receives packets from input port 0 and input port 1 like Fig. 2.12a, two slots are needed. In FiC, when a user logic sends the same data to the user logic on the other multiple boards, broadcast or multicast can relax the injection bottlenecks, but it does not mitigate the ejection bottlenecks.

In order to relax the ejection bottleneck, we propose a multi-ejection STDM switch. Figure 2.12b shows the concept of multi-ejection STDM switch. Compared to the base STDM switch (Fig. 2.3), it provides multiple ejection ports, and each port can eject packets from a certain input port [5, 18]. For example, ejection port 1 ejects packets from input port 1. By using the multi-ejection switch, the number of slots can be reduced like Fig. 2.12b. Furthermore, we provided an ejection table to avoid receiving unwanted packets. Its entry is provided for each slot, and it determines whether the ejection port ejects a packet or not in the slot.

By a combination of the link aggregation and multi-ejection switches, the communication performance was improved up to 7.50 times with small additional resources. Although the performance of the FFT with the highest communication ratio was not able to enhance by using multiple boards when a lane was used, 1.99

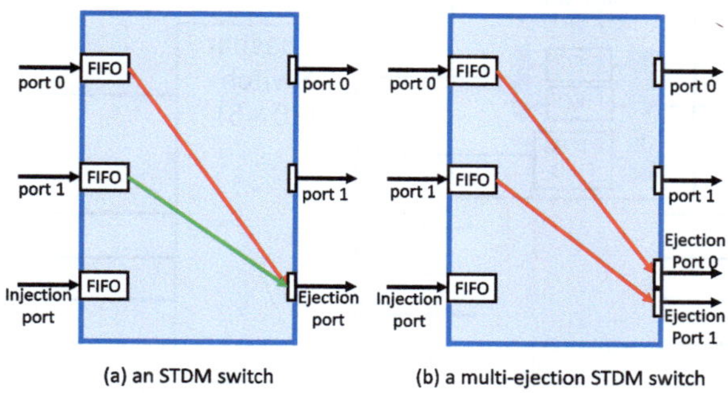

Fig. 2.12 The ejection bottleneck and elimination of it by a multi-ejection switch. (**a**) An STDM switch. (**b**) A multi-ejection STDM switch

times performance improvement was achieved by using eight boards with four lanes and multi-ejection switches.

4.2 Hybrid Router

In packet switching, packets are divided into flits, which are the basic units of transmission. When a header flit arrives, a packet switching router determines the output port from the destination address in the flit and then reserves the output virtual channels and slots of the crossbar switch until a tail flit has arrived. In a packet switching network, each router performs routing and arbitration independently, and the performance is the best effort. Compared to STDM, the packet switching performance is superior when the link utilization is low. However, when the link utilization increases, congestion occurs, and the latency of each packet is stretched. By using STDM and packet switching appropriately, our hybrid router can exploit the advantages of both switching methods.

Figure 2.13 presents a block diagram of our hybrid router. It consists of a packet switching module and an STDM module. Both modules share the crossbar switch and I/O ports. The packet switching module can send a flit when the slot of the STDM is empty even if the slot is assigned to a certain communication for STDM. The packet switching module is based on an open-source NoC router [19] that is written in Verilog HDL. We modify the design to act properly on the FiC system and to support STDM simultaneously. It is a three-stage pipelined router that is composed of the routing computation stage, virtual channel/switch allocation stage,

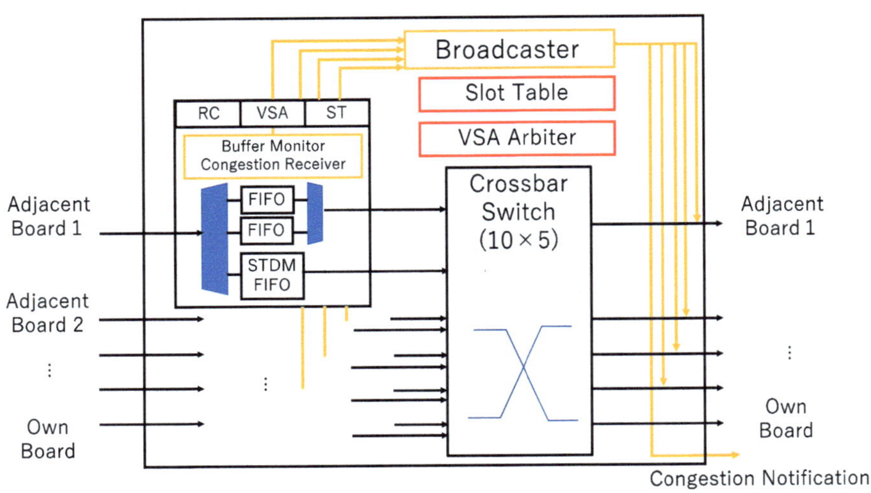

Fig. 2.13 The diagram of our hybrid router

and switch traversal stage. To suit FiC, the bit length of each flit is set to 170 bits (including a 1-bit data valid signal), and I/O is performed every two clock cycles. Note that our router body, including the routing and arbitration, operates in every cycle. Our hybrid router adopts a packet format with a valid field (1-bit), destination ID field (16-bit), type field (3-bit), virtual channel field (1-bit), acknowledge field (4-bit), congestion signal field (1-bit), and payload field (144-bit). The congestion signal field is used to enable a dynamic communication mode switch mechanism. FiC uses a 2D torus. Our router has five I/O ports, and two virtual channels are provided to avoid deadlocks. We adopt the DOR (XY routing) for the shortest-path deadlock-free routing.

Furthermore, to improve each link utilization under a low network traffic load, we propose a dynamic communication switching algorithm. In our algorithm, each router monitors the network load metrics, and according to the metrics, timing-critical tasks select the STDM according to the metrics only when congestion occurs. This can achieve both QoS guarantee and efficient utilization of each link with a small resource overhead. In our evaluation, the dynamic algorithm was up to 24.6% faster on the execution time with a high network load compared to the packet switching on a real multi-FPGA system with 24 boards.

4.3 MPI Library

To facilitate application development for the M-KUBOS cluster, we developed the parallel computing library "FiCMPI" using Vivado HLS, which is available for FPGA application development using C/C + + description in high-level synthesis (HLS).

FiCMPI is a parallel library for the M-KUBOS cluster that implements a subset of the parallel computing API standard, MPI (Message Passing Interface). It is intended to make it easier to port C<Dmathdollar>/C + +-described applications developed using the MPI API, which is commonly used in scientific and engineering applications, to the M-KUBOS cluster with minimal effort. The implementation of FiCMPI, synthesized as FPGA logic circuits through Vivado HLS (a vendor tool for HDL synthesis), imposes unavoidable constraints on hardware resource usage and HLS execution models. However, efforts have been made to ensure that application developers can use it with a familiar MPI-like experience.

Table 2.7 lists the APIs currently supported by FiCMPI. It covers commonly used primitives for one-to-one communication and collective communication. However, at present, only blocking APIs are supported for all these primitives.

Here, to evaluate the performance scalability of the M-KUBOS cluster, we used fic_himenoBMT, an MPI-based Himeno Benchmark ported to the M-KUBOS cluster using FiCMPI, to assess the system's performance scalability as the number of nodes changes. Figure 2.14 shows the performance of fic_himenoBMT measured while varying the number of nodes from 1 to 6. The bar graph corresponds to MFLOPS performance represented on the left vertical axis, while the line graph

Table 2.7 APIs supported by FiCMPI

Type	API	Function
Program control	MPI_Init	MPI initialize
	MPI_Finalize	MPI finalize
	MPI_Comm_rank	Get self-process rank
	MPI_Comm_size	Get number of all processes
1-to-1 data transfer	MPI_Send	P2P data send (blocking)
	MPI_Recv	P2P data receive (blocking)
	MPI_Recvany	P2P Any data receive (blocking)
Group transfer	MPI_Bcast	Broadcast
	MPI_Scatter	Scatter
	MPI_Gather	Gather
	MPI_Allgather	All gather
	MPI_Reduce	Reduction
	MPI_Allreduce	All reduction
	MPI_Alltoall	All-to-all data transfer
Other	MPI_Type_vector	Type vector definition
	MPI_Cart_create	Cartesian topology definition
	MPI_Cart_get	Cartesian coordinate retrieval
	MPI_Cart_shift	Cartesian coordinate shift

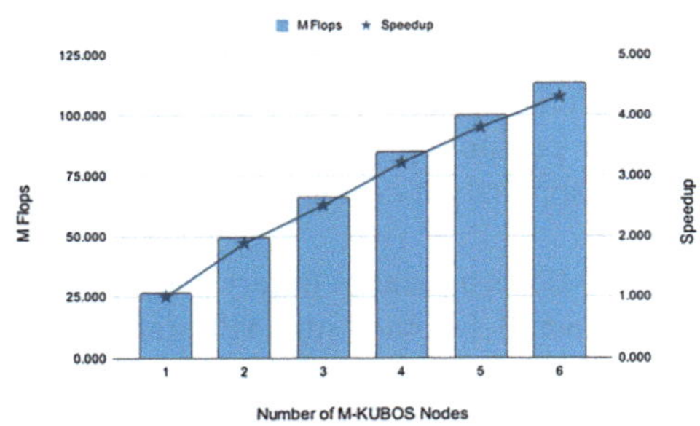

Fig. 2.14 The performance of fic_himenoBMT on M-KUBOS cluster

represents the performance increase ratio relative to a single node, as shown on the right vertical axis. The horizontal axis represents the number of execution nodes.

For a single M-KUBOS node, the performance is 26.29 MFLOPS, which is not particularly noteworthy compared to the reference performance of 82.84 MFLOPS for the Himeno benchmark running on a Pentium III 600 MHz processor. However, considering that M-KUBOS operates at 100 MHz—a sixfold frequency difference—it can be deemed reasonable performance, especially given that no specific FPGA optimizations were applied in this evaluation.

4.4 IP/Ether over GTH

On the IoT platform, all nodes are connected to the Internet, so the nodes are TCP/IP reachable. M-KUBOS has the 1 Gbps Ethernet, and the PL part can communicate another board's PL part through PS. Sometimes PL part wants to communicate with another PL part directly through GTH link. If a hardware TCP/IP protocol stack exists, IP over GTH is available, but a hardware implementation of TCP/IP is hard. So, we propose an easy way IP/Ether over GTH with tun/tap virtual device of Unix OS [7].

Usually, tun/tap virtual device is used for VPN (virtual private network) implementation. Tun/tap device is shown, such as ethX in kernel virtual NIC bound to virtual character device as virtual serial line, like /dev/net/tun. When PS program writes IP or Ethernet packet to the character device, the packets are sent to a virtual NIC tun device is L3 virtual NIC, and tap is L2.

The basic structure of IP/Ether over GTH is shown in Fig. 2.15. For VPN implementation, two machines are connected with TCP/IP socket. IP/Ether packets are sent and received through the socket; these packets are read and written to a tun/tap. Tun/tap virtual NIC has IP address; thus all commands for IP/Ether can treat Tun/tap virtual device. For example, "ip," "ifconfig," and "route" commands can be used for tun/tap. Bridge commands like "brctl" can set a tap device.

General communication line can be used to connect two machines. GTH is also available. No device driver is required for TCP/IP.

By using the M-KUBOS cluster, the performance is evaluated. Table 2.8 shows the time to transfer 10 MB data by ICMP (Internet Control Message Protocol), TCP, and UDP (User Datagram Protocol) echo including 1 KB Ether for comparison. The performance of IP/GTH is lower than that of 1 Gb Ether. This result is due to the software flow control overhead. The hardware flow control is currently under development to increase the total performance.

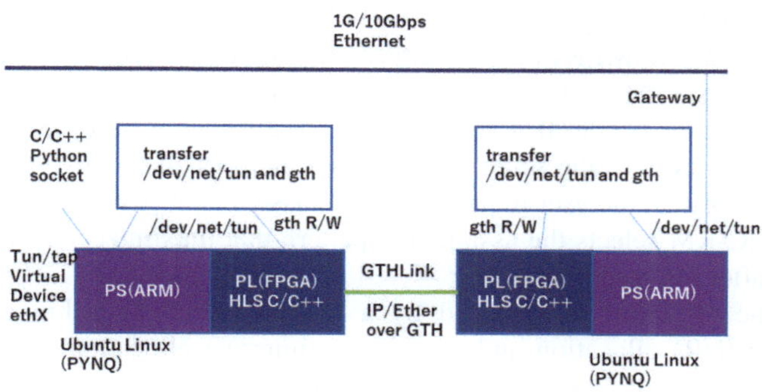

Fig. 2.15 Basic structure of IP/Ether over GTH

Table 2.8 Time for 10 MB data transfer (msec)

Comm. protocol	1Gb Ether	GTH
ICMP	0.159	0.205
TCP	1940	4837
UDP	789	2152

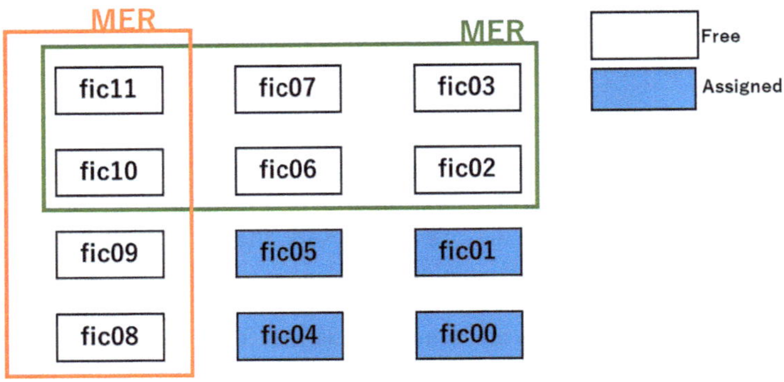

Fig. 2.16 An example of MERs

4.5 Multi-tenant Resource Management

For sharing an M-KUBOS cluster with multiple users, the multi-tenant resource management system is needed. FiC-RM [20] is developed for this purpose.

In FiC-RM, the board assignment is decided so that the shape of each slice is rectangular. Therefore, it is necessary to find a rectangular free space. FiC-RM finds a rectangular free space using the Scan Line Algorithm (SLA) proposed in [21]. By using SLA, Maximum Empty Rectangles (MERs) can be found. MER is an empty rectangle that cannot be entirely covered by any other rectangle. For example, two MERs are depicted in Fig. 2.16.

FiC-RM can find MERs that are larger than the user's request. Then, in order from the smallest MER, they are checked to see if the requested slice can be assigned in the form of a rectangle. If it can be assigned, the slice is assigned in the MER. If multiple MERs have the same size, or if the number of boards included in the MER is greater than the user request, multiple assignment patterns are possible. In this case, FiC-RM selects the assignment pattern that maximizes the size of MER remaining after the assignment.

Later, the algorithm is replaced by a more flexible algorithm called Non-Rectangular (NR) allocation and attempts to improve allocation efficiency [22]. NR allocation provides the hybrid method, which chooses column-major allocation and row-major allocation. By using this proposed hybrid method, we have made it possible to allocate non-rectangular areas while maintaining contiguous regions by considering the board's position after allocation. Simulation results show a 42%

improvement in the total waiting time compared to the conventional method. More sophisticated resource management is described in Chap. 3.

4.6 Power Model

Power models for FPGA clusters have not been extensively studied thus far. We conducted detailed power measurements for real applications on FiC and the M-KUBOS cluster and proposed a power model.

Similar to other common FPGA boards, M-KUBOS mainly uses three types of voltage supplies, which are converted from a 12-V DC power supply for the board: 0.85 V for the chip core (*Core*), 1.85 V for the chip I/O (*I/O*), and 1.2 V for the DDR-4 memory modules (*DRAM*). To measure the power consumption of each element without disturbing the operation, we replaced the original DC/DC converters with Analog Devices (ADs) LTM4677 which performs the monitoring functions and DC/DC conversion. They are connected with the Windows PC through an AD DC2086A Powered Programming Adapter and through USB to a DC1613A I^2C controller, as illustrated in Fig. 2.17. The AD software LTPowerPlay captures all data and saves it into files in CSV format so as to be analyzed later. By using UltraVNC, the data can be remotely accessed from the server of the M-KUBOS cluster, and a dedicated Python program is provided for analysis.

The detail is shown in [23]. We can estimate the power of a developed application with 15% error in the stage of design. By using the model, the optimization of the cluster structure can be investigated. The method is described in [24].

Fig. 2.17 The M-KUBOS power monitoring tool

5 Related Work

5.1 FPGAs in the Cloud

Although a lot of FPGA clusters have been reported, most of them are used in the cloud.

Microsoft develops the FPGA-based cloud acceleration platform, Catapult [25], to fulfill the demand of data center workloads like high computational capabilities and power efficiency [25]. Each node of the Catapult cluster consists of two Intel Xeon Sandy Bridge CPUs and an Altera Stratix V FPGA. The upgraded version is Catapult v2 [26], and the further advanced version focused on AI is BrainWave [27]. These Microsoft projects demonstrated that FPGA clusters effectively function as accelerators for Bing and AI in the cloud.

Although they are customized for specific applications, some cloud enables to use FPGAs in the cloud for general users. Amazon Web Services (AWS) provides EC2 F1 instances, which are specialized to realize the custom hardware accelerations using FPGAs [28]. The F1 instances offer up to eight AMD Virtex UltraScale+ VU9P-based FPGAs. Alibaba Cloud also provides FPGA-based instances as f3, FPGA-accelerated compute-optimized instance family [29].

5.2 Stand-Alone High-Performance FPGA Clusters

cloudFPGA is developed in the IBM Research Zurich [30]. It is a stand-alone cluster of only FPGAs, not a conventional combination of CPUs and FPGAs. 32 FPGAs are installed into a single carrier board (sled), and two sleds are housed in a 2U size chassis; thus, 64 FPGAs are mounted on a single chassis [31]. Cygnus is a heterogeneous supercomputer system that combines CPUs, GPUs, and FPGAs, which is developed by the Center for Computational Sciences (CCS), University of Tsukuba [32]. RIKEN proposes an FPGA-based Elastic and Scalable FPGA cluster System for high-PErformance Reconfigurable computing (ESSPER) as an accelerator to extend the Japanese flagship supercomputer, Fugaku [33]. EuroEXA has developed an FPGA cluster for HPC funded by EU Horizon 2020. Each blade of EuroEXA consists of 16 nodes, and each node has a Zynq and a Virtex UltraScale+ FPGA [34]. 32 blades are installed in this system, so 512 FPGAs are mounted in total. Their goal is to establish FPGA supercomputers and too expensive for MEC.

Nanyang Technological University's Zedwulf, a cluster using the Xilinx Z7020, which is a kind of Zynq-type FPGA board [35] is a similar approach to us. However, the FPGA is much smaller than ours, and it does not provide direct interconnection networks between FPGA boards.

6 Conclusion of This Section

This section shows various technologies for stand-alone FPGA clusters developed in this project. The M-KUBOS board was commercialized and has been used widely in this project. Our project has a common research base, and it has been extremely important for technical discussion between groups.

References

1. K. Musha, T. Kudoh, H. Amano, Deep learning on high performance FPGA switching boards: flow-in-cloud, in *Applied Reconfigurable Computing* (2018), pp. 43–54
2. Samtec Inc., *Micro Flyover On-Board Optical Engine, Firefly*. https://www.samtec.com/optics/optical-cable/mid-board/firefly. Accessed 21 Dec 2020
3. V.R. Manfrinato et. al., Determining the resolution limits of electron-beam lithography: direct measurement of the point-spread function. ACS Nano Lett. **14**(8), 4406–4412 (2014)
4. K. Azegami et al., A STDM (static time division multiplexing) switch on a multi-FPGA system, in *2019 IEEE 13th International Symposium on Embedded Multicore/Many-Core Systems-on-Chip (MCSoC)* (2019), pp. 328–333. https://doi.org/10.1109/MCSoC.2019.00053
5. K. Ito et al., Improving the performance of circuit-switched inter-connection network for a multi-FPGA system. IEICE Trans. Inform. Syst. **E104**(D.12), 2029–2039 (2021). https://doi.org/10.1587/transinf.2021PAP0002
6. Y. Hu, M. Koibuchi, Optimizing slot utilization and network topology for communication pattern on circuit-switched parallel computing systems. IEICE Trans. Inform. Syst. **E102**(D.2), 247–260 (2019). https://doi.org/10.1587/transinf.2018EDP7225
7. T. Inage et al., M-KUBOS/PYNQ cluster for multi-access edge computing, in *2021 Ninth International Symposium on Computing and Networking (CANDAR)* (2021), pp. 95–101. https://doi.org/10.1109/CANDAR53791.2021.00020
8. SCHEDMD, *Slurm Workload Manager*. https://slurm.schedmd.com/documentation.htm
9. Y. Umuroglu et al., FINN: a framework for fast, scalable binarized neural network inference, in *Proceedings of the 2017 ACM/SIGDA International Symposium on Field-Programmable Gate Arrays*. FPGA '17 (Association for Computing Machinery, Monterey, 2017), pp. 65–74. ISBN: 9781450343541. https://doi.org/10.1145/3020078.3021744
10. Y. Sun, H. Amano, FiC-RNN: a multi-FPGA acceleration framework for deep recurrent neural networks. IEICE Trans. Inform. Syst. **E103**(D.12), 2457–2462 (2020). https://doi.org/10.1587/transinf.2020PAP0003
11. Y. Yamauchi et al., Horizontal division of deep learning applications with all-to-all communication on a multi-FPGA system, in *2020 Eighth International Symposium on Computing and Networking Workshops (CANDARW)* (2020), pp. 277–281. https://doi.org/10.1109/CANDARW51189.2020.00060
12. Y. Fukushima, K. Iizuka, H. Amano, Parallel implementation of CNN on multi-FPGA cluster, in *2021 IEEE 14th International Symposium on Embedded Multicore/Many-core Systems-on-Chip (MCSoC)* (2021), pp. 77–83. https://doi.org/10.1109/MCSoC51149.2021.00019
13. Y. Fukushima, K. Iizuka, H. Amano, Parallel implementation of CNN on multi-FPGA cluster. IEICE Trans. Inform. Syst. **E106**(D.7), 1198–1208 (2023). https://doi.org/10.1587/transinf.2022EDP7175
14. M.M. Imdad Ullah, A. B. Ahmed, H. Amano, Implementation of FM-index based pattern search on a multi-FPGA system, in *Applied Reconfigurable Computing. Architectures, Tools, and Applications*, ed. by F. Rincón et al (Springer, Cham, 2020), pp. 376–391. ISBN: 978-3-030-44534-8. https://doi.org/10.1007/978-3-030-44534-8_28

15. Y.J. Yan et al., Resource-saving FPGA implementation of the satisfiability problem solver: AmoebaSATslim, in *2021 International Conference on Field-Programmable Technology (ICFPT)* (2021), pp. 1–5. https://doi.org/10.1109/ICFPT52863.2021.9609882
16. A.H.N. Nguyen, Y. Hara-Azumi, An FPGA-based stochastic SAT solver leveraging inter-variable dependencies, in *2021 31st International Conference on Field-Programmable Logic and Applications (FPL)* (2021), pp. 179–184. https://doi.org/10.1109/FPL53798.2021.00037
17. K. Ito et al., Implementing a multi-ejection switch and making the use of multiple lanes in a circuit-switched multi-FPGA system, in *2020 Eighth International Symposium on Computing and Networking Workshops (CANDARW)* (2020), pp. 211–217. https://doi.org/10.1109/CANDARW51189.2020.00049
18. K. Ito, R. Yasudo, H. Amano, Optimizing application mapping for multi-FPGA systems with multi-ejection STDM switches, in *2022 32nd International Conference on Field-Programmable Logic and Applications (FPL)* (2022), pp. 143–147. https://doi.org/10.1109/FPL57034.2022.00032
19. H. Matsutani et al., Prediction router: yet another low latency on-chip router architecture. in *2009 IEEE 15th International Symposium on High Performance Computer Architecture* (2009), pp. 367–378. https://doi.org/10.1109/HPCA.2009.4798274
20. M. Yamakura et al., A multi-tenant resource management system for multi-FPGA systems. IEICE Trans. Inform. Syst. **E104**(D.12), 2078–2088 (2021). https://doi.org/10.1587/transinf.2021PAP0005
21. J. Cui et al., An efficient algorithm for online management of 2D area of partially reconfigurable FPGAs, in *2007 Design, Automation Test in Europe Conference Exhibition* (2007), pp. 1–6. https://doi.org/10.1109/DATE.2007.364579
22. T. Inage, K. Iizuka, H. Amano, Board allocation algorithm for the resource management system of FiC, in *2023 IEEE 16th International Symposium on Embedded Multicore/Many-core Systems-on-Chip (MCSoC)* (2023), pp. 218–224. https://doi.org/10.1109/MCSoC60832.2023.00040
23. K. Iizuka et al., Power analysis and power modeling of directly-connected FPGA clusters. IEICE Trans. Inform. Syst. **E106-D**, 1997–2005 (2023).
24. K. Iizuka et al., Power optimized design framework for FPGA clusters. IPSJ Trans. Syst. LSI Design Methodol. **17**, 77–86 (2024)
25. A. Putnam et al., A reconfigurable fabric for accelerating large-scale datacenter services, in *2014 ACM/IEEE 41st International Symposium on Computer Architecture (ISCA)* (2014), pp. 13–24. https://doi.org/10.1109/ISCA.2014.6853195
26. A.M. Caulfield et al., A cloud-scale acceleration architecture, in *2016 49th Annual IEEE/ACM International Symposium on Microarchitecture (MICRO)* (2016), pp. 1–13. https://doi.org/10.1109/MICRO.2016.7783710
27. J. Fowers et al., A configurable cloud-scale DNN processor for real-time AI, in *2018 ACM/IEEE 45th Annual International Symposium on Computer Architecture (ISCA)* (2018), pp. 1–14. https://doi.org/10.1109/ISCA.2018.00012
28. Amazon Web Services, *Amazon EC2 F1 Instances*. https://aws.amazon.com/ec2/instance-types/f1/. Accessed 2 Mar 2024
29. Alibaba Cloud, *Compute Optimized Type Family with FPGA, Overview*. https://www.alibabacloud.com/help/en/ecs/user-guide/overview-39. Accessed 2 Mar 2024
30. IBM Research Europe—Zurich, *cloudFPGA*. https://research.ibm.com/projects/cloudfpga. Accessed 2 Mar 2024
31. F. Abel et al., An FPGA platform for hyperscalers, in *2017 IEEE 25th Annual Symposium on High-Performance Interconnects (HOTI)* (2017), pp. 29–32. https://doi.org/10.1109/HOTI.2017.13
32. T. Boku et al., Cygnus—world first multihybrid accelerated cluster with GPU and FPGA coupling, in *Workshop Proceedings of the 51st International Conference on Parallel Processing*. ICPP Workshops '22 (Association for Computing Machinery, Bordeaux, 2023). ISBN: 9781450394451. https://doi.org/10.1145/3547276.3548629

33. K. Sano et al., ESSPER: elastic and scalable FPGA-cluster system for high-performance reconfigurable computing with supercomputer Fugaku, in *Proceedings of the International Conference on High Performance Computing in Asia-Pacific Region*. HPCAsia '23 (Association for Computing Machinery, New York, 2023), pp. 140–150. ISBN: 9781450398053. https://doi.org/10.1145/3578178.3579341
34. EuroEXA, *EuroEXA*. https://www.euroexa.eu/. Accessed 2 Mar 2024
35. P. Moorthy, N. Kapre, Zedwulf: power-performance tradeoffs of a 32-node Zynq SoC cluster, in *2015 IEEE 23rd Annual International Symposium on Field-Programmable Custom Computing Machines* (2015), pp. 68–75.

Chapter 3
Middleware for MEC

Midori Sugaya, Takeshi Ohkawa, Mikiko Sato, and Yanzhi Li

1 Introduction

In recent years, with the rapid development of network connectivity technologies such as WLAN and cellular networks, devices with sensors, processing ability, and other technologies have become possible to connect and exchange data with other devices over the Internet or local networks. The technology for connecting devices to networks is generally called the IoT (Internet of Things). More and more IoT devices are connected to the network and making the industry, and our daily lives become convenient.

However, increasing the number of IoT devices causes a large amount of traffic to the cloud-based server, which will slow down the response time of the cloud-based architecture. Since the cloud-based architecture is generally located in a remote place, and the massive data tends to occupy the traffic, it is difficult to connect it without delay. Edge computing is one of the methods to solve the traffic influx problem in cloud-based architecture. It suggests placing an edge server closer to the IoT devices and serving a limited area with a limited number of IoT devices. Compared to the cloud server, the shorter distance from the IoT device to the edge

M. Sugaya (✉) · Y. Li
Department of Engineering, School of Computer and Science and Engineering, Shibaura Institute of Technology, Tokyo, Japan
e-mail: doly@shibaura-it.ac.jp; nb22509@shibaura-it.ac.jp

T. Ohkawa
Research and Education Institute for Semiconductors and Informatics, Kumamoto University, Kumamoto, Japan
e-mail: ohkawa-takeshi@kumamoto-u.ac.jp

M. Sato
Department of Information and Telecommunication Engineering, Tokai University, Tokyo, Japan
e-mail: mikiko.sato@tokai.ac.jp

H. Amano (ed.), *Computation Platforms for Multi-access Edge Computing*,
https://doi.org/10.1007/978-981-96-8935-4_3

server results in shorter latency. Since edge computing offers lower latency and higher throughput, it is suitable for IoT devices to offload their computing to.

We consider the application of IoT devices that connect to the edge server. Today, the variety of applications has various functions. With the rapid revolution of deep learning algorithms and other AI technologies, devices require these technologies in a smarter interactive environment. However, algorithms such as object detection or generative AI require a large scale of computational resources. With benefit from the benefit on communication, IoT devices can offload the heavy computing tasks to other computers, such as other IoT devices or edge servers. However, there are some problems with computational resource management. The first problem is how to detect a computer with sufficient capability. The second problem is how to determine which computer to execute the computing task on.

Unlike cloud servers, there are no de facto architectures for IoT devices, which is the reason for the first problem. IoT devices may have different CPU architectures (e.g., ARM, RISC-V, etc.) and may have different heterogeneous accelerators (e.g., embedded GPU, FPGA, NPU, etc.) integrated. As a result, it is unlikely that the distribution of binary files for execution will work. This problem can be avoided by architecture-based matching algorithms or an extra hardware abstract layer. A hardware-architecture-based matching algorithm, such as a tag-based matching algorithm, is likely to select the most likely machine to execute the program. However, it may fail to match if there is no possible machine available. An extra hardware abstract layer, such as a virtual machine, may solve the problem that no machine is capable of executing the program. For IoT devices, a virtual machine for some kind of physical CPUs may be very heavy, but it is possible to use a virtual machine program, such as JVM or LuaVM, to port the program to multiple platforms.

Another main problem is the scheduling problem. Using a message broker to share data between IoT devices is a general way to develop the IoT ecosystem. In this architecture, sending the calculation task to the message broker may be the simplest way to implement. However, message brokers will send the task to other machines but will not decide which machine should execute the task. As a result, a single task may be executed by multiple machines, or no machine will chain to execute the task. A job scheduler is a way to solve this problem. The job scheduler will receive all the task (or a job) execution requests and assign the task to a machine to be executed.

2 Related Work

1. Computing Clusters
   Computing clusters combine the power of multiple computers. Key types include grid computing, HPC (high-performance computing), and HTC (high-performance cluster). In the 1990s, the GIMPS project and SETI@home used grid computing, connecting computers worldwide to share unused processing

power. The BOINC project later adopted this model, which now involves more than 700,000 computers. Grid computing handles large-scale data processing but has high latency and requires customized software for different architectures, increasing the development burden. HPC clusters combine computers into a virtual supercomputer, with nodes located close to enable frequent communication. This reduces grid computing's communication delays but limits flexibility and external usage, like with IoT devices. Slurm is a popular middleware for building HPC clusters. Load-balancing clusters, used in web services, distribute requests across multiple computers via a load balancer. They handle a large number of light tasks efficiently but struggle with heavier computational loads.

2. Serverless computing
Serverless computing is a cloud computing method. To understand it, we first look at cloud computing, which provides services such as servers, storage, and databases through the Internet. Cloud computing is categorized as IaaS and PaaS. IaaS (infrastructure as a service) provides the infrastructure (servers, storage, networks) needed to build systems, offering flexibility for developers. Examples include Amazon Elastic Compute Cloud and Google Cloud Compute Engine. PaaS (platform as a service) offers a platform for application development, including operating systems and databases, reducing the operational burden by managing both hardware and software required for development. FaaS (function as a service) is another category of cloud computing, in which server processes and runtime libraries that are managed by the developer in PaaS are also managed by the cloud service provider. Since the application developer does not manage the server in FaaS, it is also called serverless computing. Amazon Lambda and Google Cloud Functions are examples of services offered in the serverless computing category. Serverless computing is a method in which the management of servers and operating systems is transferred from the developer to the cloud service operator. This is considered to have advantages for both developers and cloud service operators. From the cloud service operator's perspective, it is possible to transfer resources that are not currently being used to other services, and this has the advantage of allowing for more flexible operation of the infrastructure system. Developers also have the advantage of being able to concentrate on application development, freeing them from the management of servers and operating systems. In addition, the system can automatically respond to sudden increases in traffic without the need to reserve resources in advance.
3. Edge computing
Edge computing is one of the approaches to reduce the traffic concentrated in the cloud. Edge computing suggests a method to process a part of requests concentrated in the cloud in locations closer to end users. Since requests are distributed, the traffic concentration in the cloud can be resolved. It also helps to improve latency and responsiveness, as processing is performed near the end user. The first successful implementation of edge computing is considered the CDN (content delivery network) [1]. Originally, CDNs only delivered static content used by websites, but later, as dynamic content increased, the cases in which dynamic content was generated by CDNs also increased. For example,

CDN providers such as Cloudflare [2] and AWS [5] have been offering services to run programs on top of the CDN infrastructure in recent years. At the same time, with the development of mobile communication networks, the number of terminals connected to the Internet is rapidly increasing. Mobile terminals often experience communication interruptions as they switch between connections to multiple base stations. To improve the user experience, low-handshaking protocols, such as QUIC, are used, while low latency, which can be provided by edge computing, is also attracting attention. In addition, 5G, which is characterized by high capacity and low latency, has recently been commercialized. The cloud computing has the problem of not being able to handle the large communication capacity due to concentrated traffic, and the problem of high latency due to the remote location of the cloud. Due to these problems, edge computing is considered important in the 5G era. This edge for mobile communications such as 5G has been standardized as multi-access edge computing [2].

3 Purpose and Proposal

3.1 Issues

When we try to integrate IoT devices with heterogeneous computing resources on the edge, there are some important requirements that we should consider: (1) For heterogeneous computing resources, they may have different execution capabilities. IoT devices, especially those in consumer electronics, generally use different hardware. That means: (2) Computers that offer computational resources may not predict which calculation procedure will be used. (3) IoT devices may have an unstable network connection (typically wireless connections), and the network connection may be disconnected during the calculation that is done. (4) Both the IoT device and the computer that offer their resources may leave the current network. Especially in smart home applications, if we use the smartphone or a notebook computer as a compute node, these nodes may join or leave the area as humans enter and exit. (5) Optionally, IoT devices may be supposed to have higher responsibilities. However, when we try to select a system that satisfies those requirements, we find the problem that the well-known systems that were developed for similar scenes do not satisfy all of those requirements. The two major problems in using existing compute clusters for IoT are considered as follows: (1) Existing HPC and grid computing managers do not abstract hardware features, limiting the types of hardware that can be used. (2) Existing cluster management systems are managed by a small number of management nodes, resulting in traffic concentration and communication delays when used over a wide area. To get all the above five requirements satisfied, and aim to solve the two management problems in the environment with heterogeneous hardware and broker the computational resources between different hardware and to achieve this goal, we propose MEC-

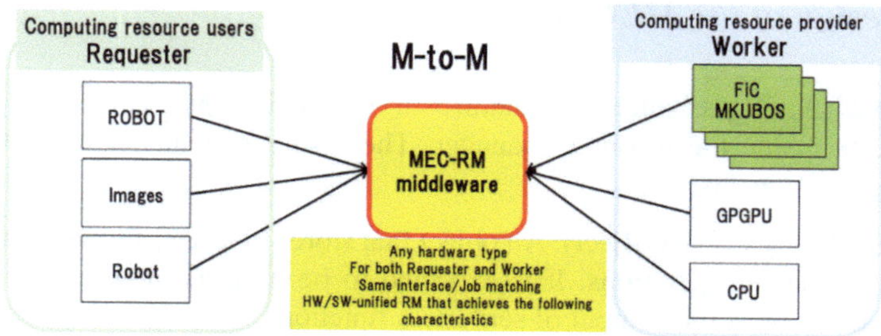

Fig. 3.1 Basic architecture of MEC-RM (MEC resource manager)

RM (multi-access edge computing resource manager) as a data analysis and processing platform that processes requests from a large number of devices, such as IoT devices. Figure 3.1 shows the basic architecture. The requester is the user of computing resources and submits computation requests to the resource manager. The worker on the other side is a computing resource provider and submits a request to the resource manager to allocate a computing task. The resource manager stands between the computing resource user (requester) and the computing resource provider (worker) and mediates computing power. Computation is requested and executed in the form of a job. From the viewpoint of the protocol sequence, the requester sends the job request by HTTP POST requests, and the resource manager (MEC-RM) finds and matches a worker by the given tag and assigns the job to the detected worker by an HTTP long-poll connection. The worker who gets the assignment will execute the given job and use the HTTP POST request to send the result back to the resource manager. The requester can use HTTP GET to poll the job status and get the results.

3.2 Proposal of the MEC-RM

Our proposed system should also have the following features: (1) A tag-matching-based scheduling algorithm to satisfy the mixed heterogeneous hardware environments. (2) Integration of serverless computing to calculate the distribution and execution of the procedure as an abstract layer of hardware. (3) A stateless transport protocol that allows long-term execution with unstable connections. (4) The IoT machine and the calculation machines can join or leave the system freely. The system should manage the status of the computation machines. (5) Stateless APIs, which allow custom congestion control algorithms for asynchronous processes to get better performance for IoT devices. We propose (Fig. 3.1).

To the best of our knowledge, there are no proposals for heterogeneous systems that transparently handle FPGA and GPGPU computing clusters deployed at the edge, so this is a new idea for managing MEC.

3.3 Design of the MEC-RM

Figure 3.2 shows the internal structure of the MEC-RM. There are multiple components inside the resource manager. The resource manager consists of the following components:

- BLOB (Binary Large OBject): An object that stores binaries, allowing the create, read, and delete operations. It is expected to be used as input/output data for calculations or as the content of serverless functions.
- Function: A serverless function object that supports CRUD operations. The contents are assumed to be the runtime required for execution and the main body of the program (reference to a BLOB object).
- Job: A job is an object that represents a computation request and supports CRUD operations. Its contents store the function to perform this calculation (reference to a function object), input/output data (reference to a BLOB object), the worker that executes this job (reference to a worker object), or the execution state.
- KV: An object for managing and storing key-value type data. Since all objects such as BLOBs, functions, and jobs are managed by IDs, information cannot be shared among jobs unless the IDs can be guessed. To avoid this limitation, a key-value type database is provided. Since KV is divided into multiple namespaces, it is necessary to share prior namespace IDs. The sharing of a namespace can be done by describing it in the contents of serverless functions.
- Scheduler: The scheduler is a component that performs job scheduling. It allocates jobs while considering the connection status (online/offline) and execution capacity (available runtime) of the worker and reschedules jobs when it detects worker departure (failure or cluster scale-down). If a worker leaves the cluster (failure or cluster scale-down), the component reschedules the job.

  MEC-RM cooperates with these multiple components to mediate computing power. Each component can be distributed across multiple machines, and the components themselves are designed to be deployable on multiple machines. This design makes MEC-RM itself a scalable system.

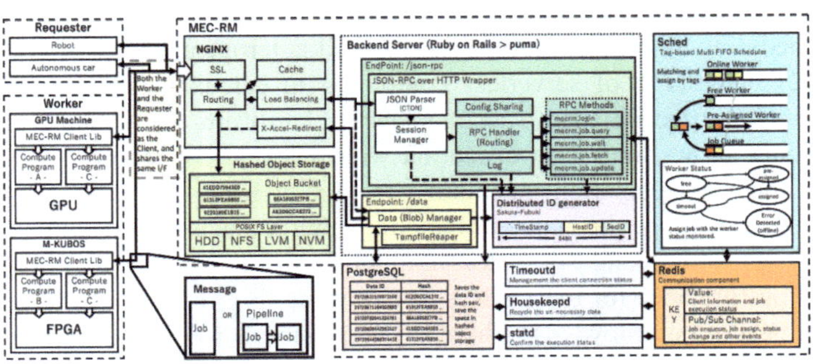

Fig. 3.2 Design of MEC-RM

4 Scheduling Method on MEC-RM

4.1 Background

In our study, we consider a scenario in which heterogeneous clusters are deployed as edge servers at base stations. Figure 3.3 illustrates this imagined scenario. In such a setup, leveraging FaaS (function as a service) offers several advantages, including flexible service usage independent of the MEC server's location. Specifically, even if a user moves to a different base station, the function (a program created by the developer) invoked by the user is executed for each request, regardless of the user's location. The functions are not limited to a single MEC cluster, allowing dynamic execution across clusters. Removing functions from the server post-execution reduces memory usage and power consumption. Moreover, this distributed approach enables the use of the service without specifying the server for execution, promoting a seamless and distributed environment for application execution (Fig. 3.3).

This section provides an overview of the functions and scheduling mechanisms of MEC-RM, along with elucidating its relationships with base stations and function as a service (FaaS). Understanding these components is crucial for understanding the operational dynamics and efficiency enhancements enabled by MEC-RM within the edge computing environment. As we described in the previous section, MEC systems in heterogeneous environments integrating high-performance accelerators such as GPGPUs and FPGAs require a resource management system that enables transparent and efficient use of these accelerators. Li proposed MEC-RM as a middleware that handles hardware transparently [4]. As shown in Fig. 3.1, MEC-RM designates robots and IoT devices as computational resource users (requestors) and provides optimal resource management functions between requestors and computational resource providers (workers). MEC-RM creates jobs and allocates workers based on computational processing requests from requestors. MEC-RM and workers are here referred to as an MEC cluster. MEC-RM provides a scheduler function that allocates workers in response to requests from multiple requestors as

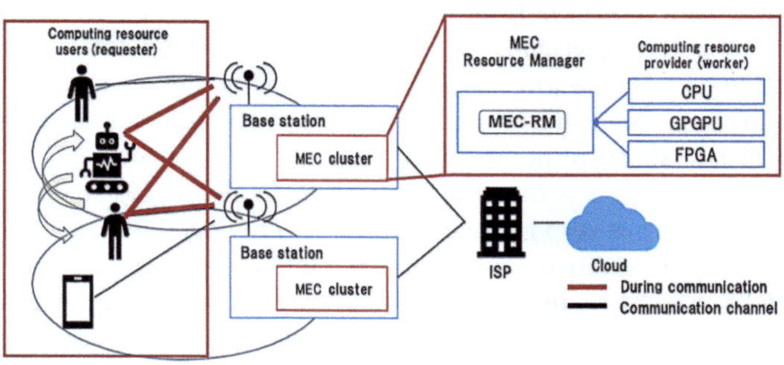

Fig. 3.3 Example of FaaS execution on an MEC cluster

shown in Fig. 3.3. This scheduler provides a basic scheduling function that allocates workers corresponding to the tag information specifying workers such as FPGAs and GPUs from requestors. When the scheduler receives a job request from a worker, it places the worker in the worker queue. In the current MEC-RM, workers request jobs to MEC-RM according to the status of their own computing resources, so the more workers in the queue, the more choices of workers that can be allocated to a job.

4.2 Issues

In the context of multi-access edge computing (MEC), the traditional approach of reserving substantial computing resources akin to cloud environments is impractical. Unlike the cloud, MEC faces constraints where allocating and occupying computing resources across any MEC cluster for running applications may not always be feasible due to communication limitations with requestors. The existing methods used to manage heterogeneous environments in the cloud are not directly applicable to MEC scenarios. This discrepancy arises from the necessity for requestors to communicate consistently with a specific server to occupy computational resources using instances and host applications on that server [9]. One solution to this problem is the use of FaaS. It has an application that consists of a program called a function, in which the developer implements a short process and a runtime to run it. Compared to existing methods of operation, FaaS is lightweight because each application is invoked by a single function. By executing a function for each request from the requestor, it is possible to run the same application on multiple MEC clusters without consuming more computing resources than necessary. This allows requestors, such as automobiles [6–8], which are expected to use MEC and may move across multiple base stations, to run any application on any MEC clusters with which they can communicate, without the need to always communicate with the same MEC cluster.

On the other hand, in FaaS, the runtime is shut down (RUNTIME SHUTDOWN) after a certain period of time from function execution (INVOKE) in order to prevent applications from constantly occupying and consuming computing resources (Fig. 3.4). Compared to a warm start, in which a running runtime is reused to execute a function, a shutdown requires runtime initialization (cold start), and its runtime startup overhead reduces responsiveness [3].

In FaaS scheduling, it is often discussed how to avoid cold starts and increase warm starts to improve responsiveness [10, 11]. Previous work, such as [12, 13], has pointed out that locality considerations can improve responsiveness by reducing the overhead caused by runtime initialization. Locality in FaaS here means the same thing as locality of reference, where the same memory address is accessed multiple times, and in this case, refers to the execution of the same function on the same server in a cluster. In the same way that the overhead can be reduced by taking into account the locality of not swapping frequently referenced data, in our proposal, by

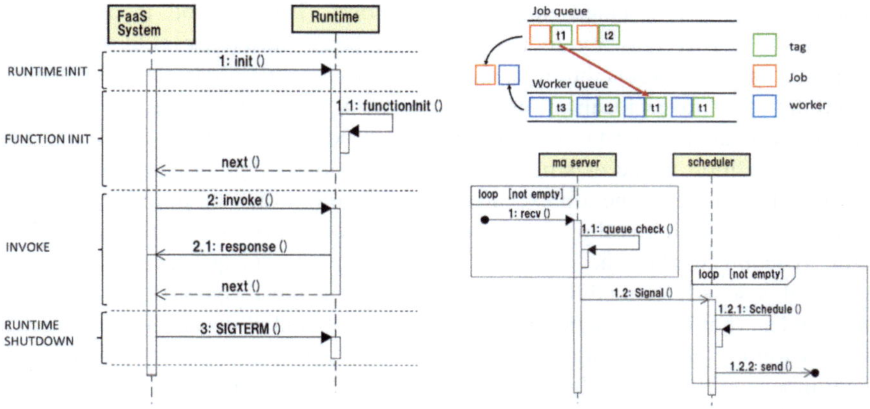

Fig. 3.4 Left: Application lifecycle in FaaS, Right upper: Locality-aware scheduler. It avoids cold starts by preferentially assigning jobs from requestors to workers that do not require initialization (i.e., each worker re-executes the function that was executed immediately before). Right lower: Schematic diagram of scheduler operation

executing the invoked function on the same server, the runtime shutdown in Fig. 3.4 can be avoided, and the probability of a warm start can be increased.

4.3 Proposed Scheduler

This study focuses on addressing the challenge of low responsiveness in FaaS-aware multi-access edge computing (MEC) systems. Our proposed solution aims to mitigate this issue by introducing a method that reduces cold starts through scheduling with careful consideration of locality. Ultimately, our goal is to significantly improve system responsiveness and overall performance within the FaaS-enabled MEC environment.

As shown in Fig. 3.4, the MEC-RM scheduler refers to the tag information (tag: t) of the job at the top of the job queue. Here, t, which represents the prefix of the tag, is a label indicating the desired worker attribute on the requestor side, such as FPGA or GPU. Here, there are three types of tags: t1, t2, and t3. Similarly, in the queue of workers, each worker is assigned a tag (t). The scheduler first checks and retrieves the tag attached to the first job. Next, it looks for a worker (t1) with a matching tag at the top of the worker queue, and when it finds a matching job (t1), it removes it from the queue. Finally, the job with the tag t1 is assigned to the worker with t1 and executed. In this way, jobs are always assigned to workers who can perform them. In contrast, we propose a locality-aware scheduler to solve the aforementioned problem.

This scheduler avoids cold starts by preferentially assigning jobs from requestors to workers that do not require initialization (i.e., each worker re-executes the function that was executed immediately before). This method is expected to improve responsiveness by reducing the overhead caused by cold starts. The proposed method is a scheduling method that takes locality into account. To achieve scheduling that takes locality into account, workers add information on the last function executed to the tag information they send to MEC-RM when requesting a job. In addition to the existing tag information, by assigning a worker to a job that matches the last function executed, the probability that the runtime is running is increased, and this is thought to reduce the probability of a cold start. If there is no matching worker, the worker at the top of the worker queue is assigned. We named this scheduler method Proposed1. However, the Proposed1 scheduler has the problem that if the function information of a job and all workers do not match, the scheduling method assigns the first worker in the worker queue and randomly determines the function to be shut down from the function information of workers waiting in the queue. As a result, functions that are called infrequently may be shut down, resulting in a higher cold start rate than functions that are called more frequently.

To solve this problem, we propose a proposed2 that takes into account the ratio of each function held by workers waiting in a queue to the total number of waiting functions (wait rate) and selects a task with a higher probability to be executed by a worker with a higher wait rate.

4.4 Evaluation

A maximum uplink communication speed of about 10 Gbps is required for 5G usage scenarios [14]. For this scenario, assuming that the application uses images of about 125 KB per image and generates 10,000 jobs per second, we created and used functions based on probability distributions created from a public dataset of Azure Functions [15]. For the evaluation, we prepared 10 functions based on a probability distribution created by extracting 50 functions from the data set in descending order of the number of calls, and numbered from 1 to 10 in descending order of call probability.

The experimental environment was constructed. The three physical machines were used for one MEC-RM (on Intel Core i7 12700kf, DDR4-3600 32 GB), and a total of 100 virtual workers (vWorker) were hosted within each machine (AMD EPYC 7543, DDR4-3200 256 GB), each running independently as a process. Each virtual worker hosts multiple runtimes for the execution of functions, and each worker can execute multiple jobs simultaneously (Fig. 3.5).

Comparing the results of all functions between the two proposed methods, Proposal 2 improves the cold start of functions that are called infrequently, allowing more workers to wait in the worker queue and more workers to be selected for a job. This is thought to reduce the number of cold starts by increasing the likelihood of selecting workers who can be warm-started during scheduling. In the experiments, the number of workers was set by assuming an application using

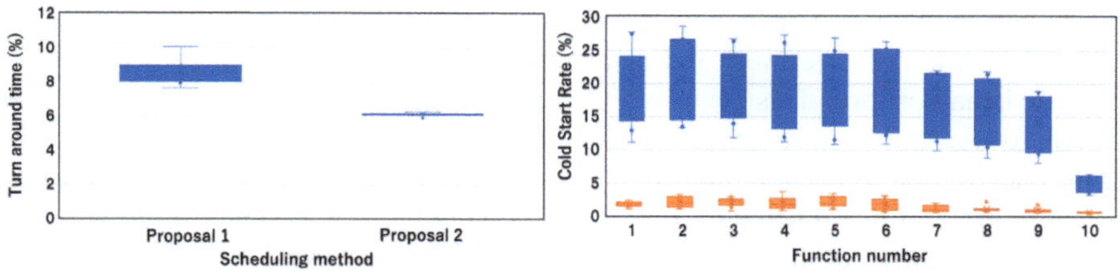

Fig. 3.5 Right: Cold start rate per function for proposal 1 and proposal 2 (blue: proposal 1, orange: proposal 2). Left: Comparison of turnaround time for the proposal 1 and proposal 2

images of approximately 125 KB for a maximum bandwidth requirement of 10 Gbps. On the other hand, in a real-world environment, bandwidth is not always used to the maximum extent, and there is a possibility of overallocating computing resources, or conversely, burst traffic that forces rapid scale-up in FaaS [16] or incorrect estimation of the expected number of jobs may cause excessive cold starts as in existing methods, resulting in poor responsiveness. Therefore, it is necessary to consider how to determine the number of worker requests for the predicted number of jobs as a future issue.

5 Application and Multi-FPGA for MEC-RM

5.1 Background

In 2022, the population over 65 years of age in Japan will exceed 28%, and the aging population in Japan is higher than in European countries [17]. Since 2000, the shortage of caregivers has continued [18], and robots and ICT (Information and Communication Technology) are required to compensate for this problem. This research aims to develop a system in which humans and robots can cooperate to provide support. This paper introduces the concept of a care support system in which robots autonomously talk to the care recipient, carry objects, guide the care recipient, and detect falls to assist the caregiver. When designing a robot for a long-term care facility or disaster relief, it is not always possible to disclose all required specifications, such as support details and persons requiring support information, in advance. For example, the following situations are not the same in each facility (nursing home):

- Number of caregivers
- Multi-access edge computing (MEC)
- Layout and structure of rooms in care facilities

Therefore, even if a caregiver support robot system can be optimally designed according to the requirements and specifications interviewed in a care facility, it faces the problem that it cannot be used sustainably under different conditions. As the situation of caregivers and residents changes, the specifications required for robotic behavior should also evolve. For example, a case may arise in which the user wants to use more robots to carry things rather than a talking robot. A system that continues to adapt to changing conditions should address the following technical issues: (1) Creating all behavior for all robots would lead to code bloat, complexity, and management complications. (2) It is too much trouble for humans to keep adjusting all of the robot's parameters and behavior optimally. In this chapter, each of the swarm robots learns the optimal action scenario while working. The behavior models and evaluations learned for each robot situation are centrally managed. This will enable a sustainable care support system that adapts autonomously to its environment.

An overview of the system under planning is shown in Fig. 3.6. Swarm robot systems, in which multiple robots perform cooperative behavior, are expected to increase efficiency through parallel work [19, 20]. The following multidisciplinary technologies are needed to realize environmentally adaptive swarm-intelligent robots based on evolutionary computation:

- Robot operating system (ROS 2) [21]
- Multi-access edge computing (MEC)
- Evolutionary particle swarm optimization (PSO) [22]
- Image processing on heterogeneous FPGAs

The MEC will be used to centrally manage the training data of each robot and provide a bird's-eye view of the behavior and status of all robots. The behavior of the entire care support robot will be optimized using a PSO-based evolutionary computation algorithm [22]. In this prototype system, we present two functions

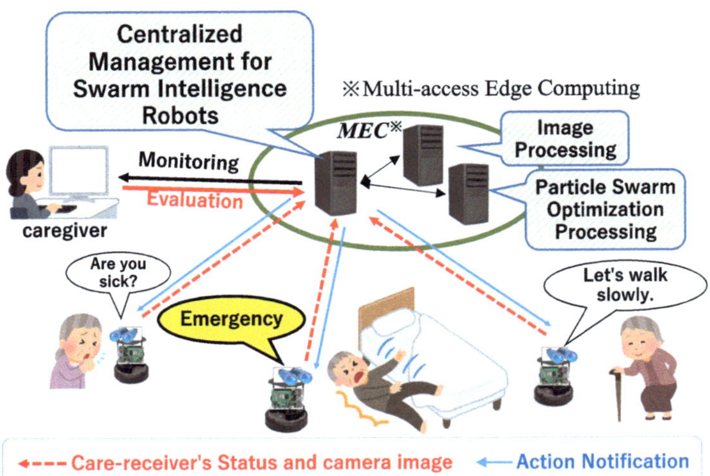

Fig. 3.6 Conceptual system of sustainable nursing care home with MEC

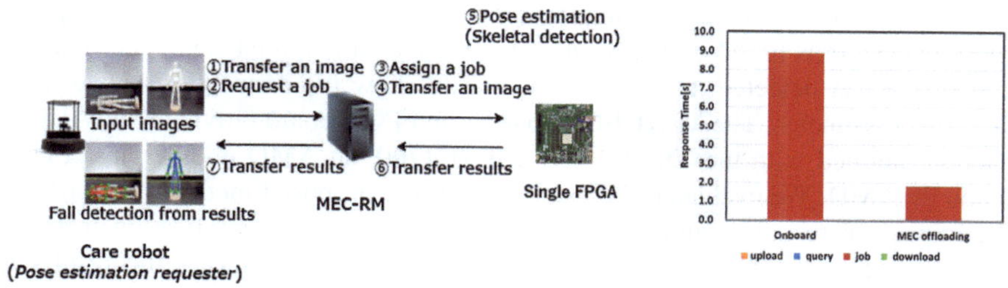

Fig. 3.7 Left: offloading system using MEC-RM (single), right: comparison of execution

implemented with the MEC-RM interface. Autonomous control of each robot, such as SLAM (simultaneous localization and mapping), is designed on the ROS (robot operating system) [23] platform. Care robots patrol the rooms of care recipients and take pictures of their posture. Pose estimation processing is computationally demanding for a low-power microcontroller in the robot system. To solve this problem, offload the pose estimation to multiple FPGAs in the MEC. The FPGAs in the MEC are predeployed in the MEC-RM. When the robot requests a pose estimation (job) from the MEC-RM, the MEC-RM assigns the process to one of the FPGAs. The robot receives the skeletal coordinates as a result of the pose estimation via the MEC-RM. As shown in Fig. 3.7, the care robot detects falls using the skeletal coordinates received from the MEC. A feature of MEC-RM is that the number and location (i.e., network address) of computing resources running in the MEC can be used without managing them on the edge device. The configuration of a homogeneous FPGA system. Furthermore, if GPGPUs or alternative types of FPGAs are deployed, edge devices can also utilize computing resources transparently via MEC-RM. In the future, MEC-RM will be able to optimize resource allocation with load-balancing and scheduling techniques that we proposed in the previous section.

As mentioned earlier, it is difficult to perform sufficient application processing with the limited computing resources on the robot. For this reason, offloading to the MEC server is expected. In this chapter, we describe the offloading via the MEC-RM middleware that we have developed. First, we will discuss the results of offloading image recognition processing, which can be expected to have general-purpose uses such as detecting the behavior and falls of the elderly, to an FPGA. In order to be able to discuss the response times of the CPU and FPGA, we will compare the results of executing the posture estimation processing on a Raspberry Pi 4B, which is a computer mounted on a robot, and executing it on an FPGA via MEC-RM. We used the M-KUBOS board [24] as the FPGA to which processing is offloaded. When the posture estimation requester (1) uploads the input image to MEC-RM and (2) requests posture estimation processing, (3) MEC-RM assigns a job to the posture estimation worker. The assigned posture estimation worker (4) downloads the input image from MEC-RM, (5) performs the posture estimation processing on the FPGA, and then (6) uploads the results to MEC-RM. When the posture estimation process is complete, the (7) posture estimation requester downloads the results. The patrol robot detects falls based on the results of posture estimation (skeletal coordinates), and if a fall is detected, it draws a red line to show the connection between the skeletal coordinates. The M-KUBOS board is an FPGA board provided by PALTEK and is equipped with the XCZU19EG-2FFVC1760, an AMD middle-class FPGA made by AMD. We evaluated the usefulness of the proposed method by clarifying the impact on response time by offloading the implemented application from the edge device to the FPGA placed in the MEC. Figure 3.7 shows the results of offloading the attitude estimation process to the FPGA on the MEC. We compared the results of running the same process on the Raspberry Pi 4B (onboard) and offloading it to the MKUBOS board using MEC-RM (MEC offloading). As a result of offloading the attitude estimation process, the response time was reduced by 79.5% compared to onboard. Since the response time for the attitude estimation process is large for onboard, the proposed method shows that the response time can be reduced by offloading the high-load process to be executed on the edge device.

5.2 Pose Estimation Process Using Multi-FPGA

In this section, we will outline a system that offloads the posture estimation process to a multi-FPGA, assuming that the detection of falls by residents of nursing care facilities is achieved using MEC. Figure 3.8 shows an overview of the multi-FPGA system and the posture estimation processing system that uses it. In this system, a patrol robot patrols the rooms of nursing home residents and uses camera images as input to detect falls by estimating posture. Since posture estimation processing is a computationally intensive process for the robot's low-performance microcontroller, posture estimation processing is offloaded to the multi-FPGA system via MEC-RM (Fig. 3.9).

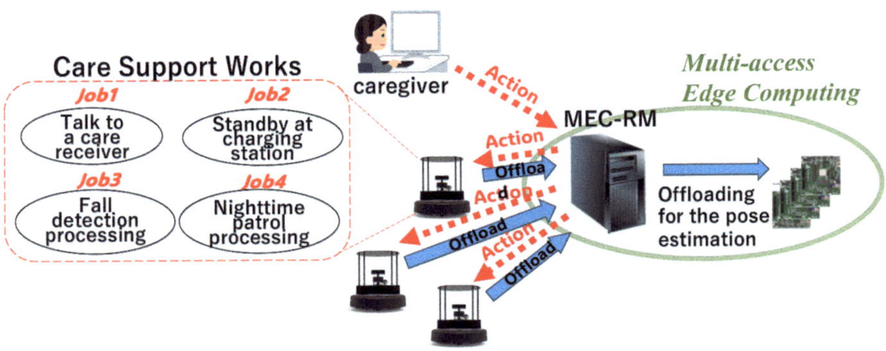

Fig. 3.8 Requirement for multiexecuting for MEC-RM on nursing care service

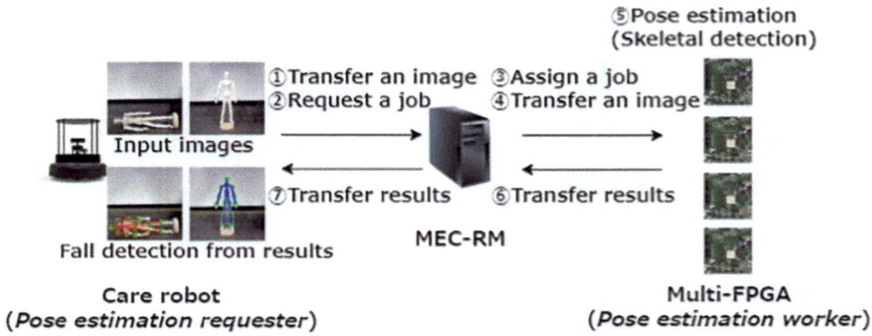

Fig. 3.9 Offloading system using MEC-RM (multi-FPGAs)

In this chapter, we use M-KUBOS as a computing resource in MEC. M-KUBOS is equipped with xczu19eg, which has the largest FPGA hardware resources among the Zynq UltraScale+ series equipped with Arm Cortex-A53 (4 cores) and is suitable for low-latency processing in MEC and AI inference. For the posture estimation process in M-KUBOS, it uses the deep learning processing unit (DPU) [25] provided by AMD Xilinx's Vitis AI [26]. An M-KUBOS can use two DPUs (DPUCZDX8G B4096 architecture) for its FPGA resources. The OpenPose model, which has been quantized to int8 precision, is used as the pose estimation model that runs on the DPU. The OpenPose class in the Vitis AI library is used for pre- and postprocessing that runs on the CPU. In preprocessing, the image size is resized to 368x368, the image is copied to the input tensor buffer, and the data are transferred from the PS (processing system) to the PL (programmable logic) memory. In the postprocessing stage, the image is copied from the PL memory to the PS output tensor buffer, and after calculating the most likely connection of the skeletal coordinate points, the skeletal coordinates are stored in a structure that holds the skeletal coordinates.

Although two cores can be installed in one M-KUBOS, in order to determine whether to use two cores when offloading from the end device as an MEC worker, we conducted a preliminary evaluation of the throughput and power efficiency of the M-KUBOS alone, and the processing time on the DPU was longer than the processing time on the CPU. This shows that the FPGA's computational resources are being fully utilized. In addition, the results of the evaluation of the throughput and power efficiency of M-KUBOS alone showed that the use of two DPU cores improves throughput and power efficiency. Based on these results, we decided to use two DPU cores in a single M-KUBOS to make the most of the computational resources inside M-KUBOS and improve performance in the implementation of the attitude estimation worker. In this chapter, we implemented the posture estimation worker so that up to two processes could be started, and each posture estimation worker would use one DPU core.

We evaluated the response time, throughput, and power efficiency using the posture estimation worker and discussed the results of comparing response time

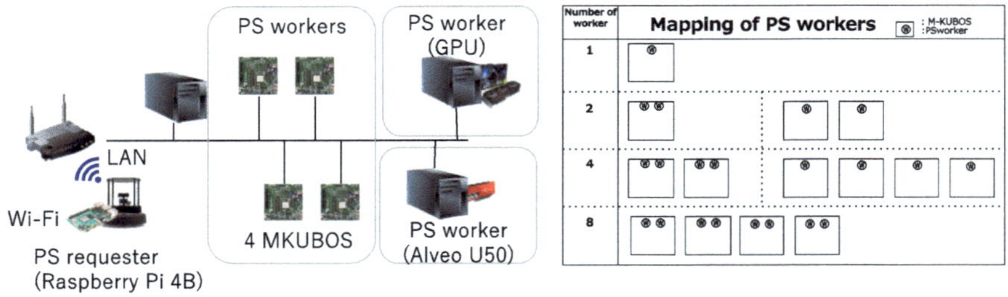

Fig. 3.10 The left figure shows the network configuration of the system used for the evaluation to offload the posture estimation process. The right figure shows the composition patterns for the posture estimation workers on the M-KUBOS

Fig. 3.11 Comparison of response time (ms) of the proposed and others

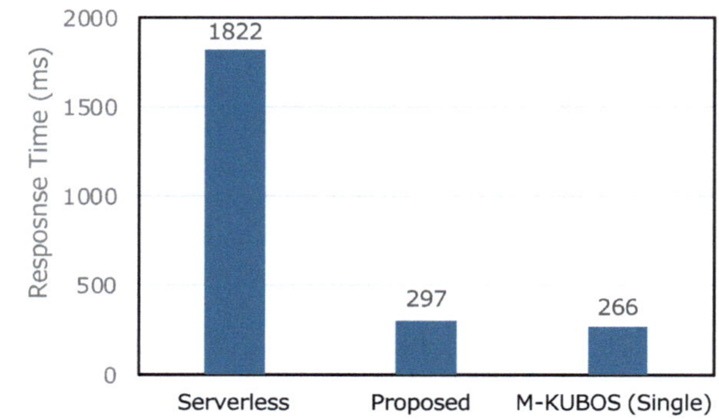

with the serverless method. First, we show the network configuration of the system used for the evaluation to offload the posture estimation process in Fig. 3.10.

In this evaluation, we use a Raspberry Pi 4B as the posture estimation requester. The four M-KUBOS machines (ARM Cortex-A53 (4 cores)) that are the posture estimation workers, AMD Xilinx Alveo U50 (Intel Core i5-12400 (12 cores)) and GPU (Intel Core i9-10940X (28 cores)), for performance comparison were connected to the LAN via a wired connection (1000BASE-T). In this evaluation, we compare the case where two posture estimation workers are configured on one M-KUBOS and the case where two posture estimation workers are configured on two M-KUBOS, respectively. In addition, the same comparison was made for the case where there were four posture estimation workers. In this section, when describing the evaluation results using multiple FPGAs, the number of posture estimation workers and the number of M-KUBOS are expressed as the "number of workers (number of units)." Figure 3.10 shows the network configuration and the worker configuration.

Figure 3.11 shows the response times for the serverless method [27] and the method of having the program resident in the M-KUBOS described in this chapter when one worker starts on an M-KUBOS and offloading is performed. The response

Table 3.1 Evaluation results for power efficiency (all measurements were taken with one machine and one worker only)

	GPU	M-KUBOS	Alveo U50
Performance (reception count/second)	15.7	3.2	9.1
Power consumption (W)	278.0	28.8	82.0
Power efficiency (performance/power consumption)	0.06	0.11	0.11

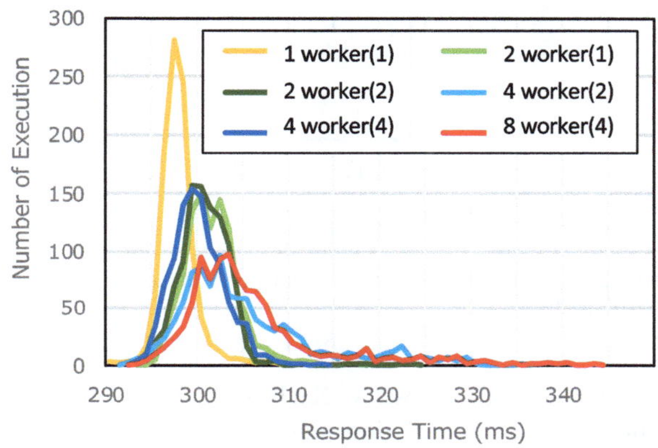

Fig. 3.12 The evaluation of correspondence between worker and DPU cores. Comparison of the response time (ms) to a different number of workers

time is the time from immediately before the input image is uploaded by the posture estimation requester to immediately after the results of the posture estimation process (skeletal coordinates) are downloaded. When the posture estimation process was offloaded to M-KUBOS using the serverless method, the response time was reduced by 83.7% to around 300ms, compared to the resident type. Furthermore, when the posture estimation process was performed using M-KUBOS alone without using MEC-RM in Fig. 3.11), the increase in response time was limited to 11.6

Table 3.1 shows the power efficiency when the posture estimation process is transferred from the Raspberry Pi 4B to M-KUBOS, GPU machine, and Alveo U50 machine. In all cases, the power efficiency was calculated from the total power consumption of the machine using a BSIDE ACM91 clamp meter. The power efficiency of M-KUBOS was 1.8 times higher than that of the GPU machine. Similarly, the Alveo U50 machine also showed a power efficiency 1.8 times higher than that of the GPU machine.

Figure 3.12 shows the response times distribution when offloading 1000 times from the posture estimation requester to M-KUBOS with each worker number of one worker (yellow), two workers (yellow green and green), four workers (light blue and blue), and eight workers (red) launched on M-KUBOS. Here, response time refers to the time from immediately before the posture estimation requester starts uploading the input image to immediately after the posture estimation processing results (skeletal coordinates) are downloaded. The response time for one worker was almost always within 300 ms. For two workers, the response time was slightly

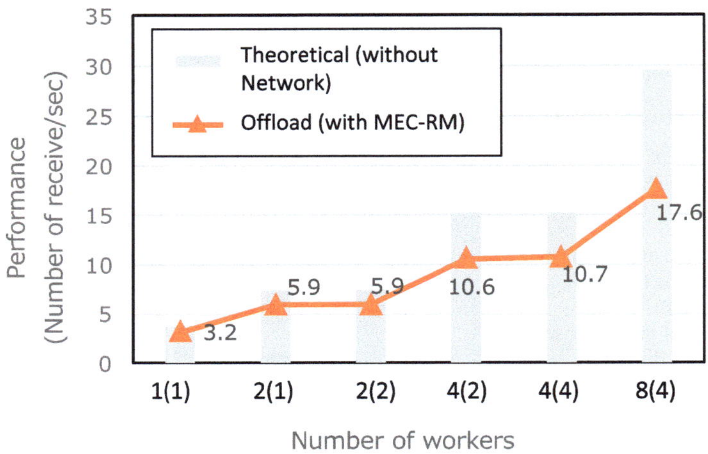

Fig. 3.13 Comparison of theoretical performance and the actual offloading performance with MEC-RM

shorter when using two M-KUBOS units. For four workers, the response time was shorter, and the variation was smaller when using four M-KUBOS units. From these results, we found that to reduce the response time and suppress variation, it is necessary to increase the number of M-KUBOS units to form a posture estimation worker.

Figure 3.13 shows the throughput when the attitude estimation process is offloaded to up to four M-KUBOS units. Throughput refers to the number of times the attitude estimation requester downloads the results of the attitude estimation process per second. The bar graph shows the ideal throughput for each number of M-KUBOS units, calculated from preliminary evaluations, and the line graph shows the actual throughput when offloading the attitude estimation process. When using one worker, throughput was 3.2 times/second, but when using two workers, it increased to 5.9 times/second, and when using eight workers, it reached 17.6 times/second. In addition, the performance of the 2- and 4-worker configurations was not greatly affected by the configuration of the posture estimation workers. Furthermore, since throughput increased almost proportionally as the number of posture estimation workers increased, it was found that MEC-RM effectively managed multiple computing resources and achieved parallel processing.

Figure 3.14 shows the power efficiency when the attitude estimation processes are offloaded to a maximum of four M-KUBOS. Compared to the power efficiency of one worker, two workers (1 unit) is 1.9 times, and four workers (two units) is 1.7 times, and the power efficiency has improved by using multiple attitude estimation workers and multiple M-KUBOS. In addition, by using two posture estimation workers on each M-KUBOS, it was possible to process with high power efficiency.

Finally, we summarize the evaluation discussion (Figs. 3.11 and 3.12). Compared to the serverless method, the resident method reduced the response time by 83.7%,

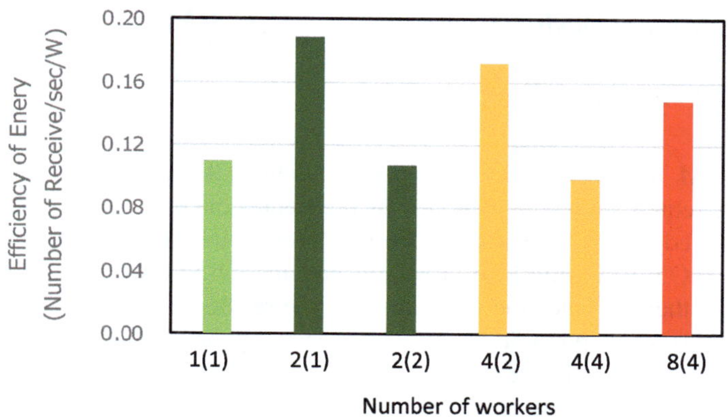

Fig. 3.14 Evaluation of power efficiency for the different number of workers

and it is believed that offloading was made possible in a practical response time. However, compared to the case where processing was performed only on M-KUBOS, the response time increased by 11. 6% when the work was offloaded to the posture estimation worker of the resident method. The main reason for this increase is the delay caused by communication with the MEC-RM. In order for the MEC-RM worker to approach the processing time of M-KUBOS, it is necessary to separate the attitude estimation processing and communication processing, and to have a system that can perform tasks such as acquiring new jobs from the MEC-RM in parallel while the estimation processing is being performed. In addition, the throughput results in Figs. 3.13 and 3.14 show that it is possible to offload with stable response times by reducing the number of MEC workers per device and using distributed computing resources with a larger number of devices. It was confirmed that offloading with high power efficiency is possible by using a configuration that makes the most of the FPGA per device (in this evaluation, two posture estimation workers were configured on one M-KUBOS). Therefore, the issue with using multiple FPGAs is choosing the optimal MEC-worker configuration according to the objective, whether it is to prioritize response time stability or power efficiency.

In the method of resident MEC worker described in this section, circuit information and processing models are loaded onto the computing resources and kept ready before the computing job is acquired. This has the advantage of being able to start computing immediately without having to perform the initialization process again when the computing job is allocated, thus reducing the response time. It is expected that a system that flexibly allocates computing resources based on demand will be provided. In addition, in the resident type method, MEC workers who have loaded specific circuit information and models continuously handle the same processing. Therefore, an implementation that switches to a different processing content during execution is expected to be provided.

6 FPGA Faas Design Example for MEC-RM

6.1 MEC-RM and Faas for FPGA

In MEC-RM [4], there are two types of clients: the requester, which makes processing requests, and the worker, which executes the processing. The requester, which is an edge device, sends a processing request to MEC-RM in the form of a job using JSON-RPC, and MEC-RM allocates the job to a worker that is waiting. The worker then executes the job and sends the results to the requester via MEC-RM. This makes it possible to offload processing on edge devices to external computing resources. However, one of the issues with the existing MEC-RM is that when offloading using an FPGA, there is an issue where overhead occurs each time a processing request is received, such as circuit reconfiguration and loading of a trained model. To solve this issue, we propose MEC-Executor, which is an extension of existing MEC-RM and is a driver for C/C + + functions for FPGAs using a scripting language.

6.2 Lua-Based C/C + + Function Calls and the Design of FaaS Middleware

In the case of FPGAs, the bottleneck in response time is caused by the need to perform initialization processes such as circuit reconfiguration and loading of pretrained image recognition models for each request. Therefore, with the aim of reducing response time and improving versatility in offloading using FPGAs in MEC, we have implemented MEC-Executor for FaaS, which calls the necessary C/C + + functions for FPGAs in response to processing requests. The processing flow is shown in Fig. 3.15. This FaaS platform is implemented by creating Lua and C/C + + libraries that call C/C + + functions using Lua and then calling them from MEC-Executor. Lua is known for its excellent performance and has been evaluated as the fastest interpreted scripting language in several benchmarks. In addition, Lua has been used to develop parts of large-scale applications, proving its practicality.

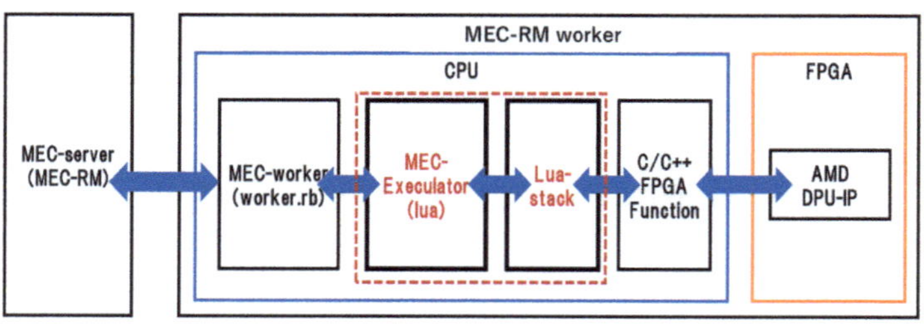

Fig. 3.15 MEC FPGA FaaS design example

First, in C/C + +, the functions to be called by Lua are registered in a table. A table is a Lua mechanism for structuring data and is an associative array. It can handle numbers and letters as keys. In the proposed method, this is used to design a system that can execute the loading of the learned model and the posture estimation process separately. MEC-Executor is called when the MEC worker receives a processing request, loads a shared library (.so file) containing C / C++ functions for the FPGA, and calls the necessary functions according to the processing request. This reduces the time required for reconfiguring the circuit and loading the pretrained model for each processing request. The implementation uses the Lua scripting language Lua. Lua is able to communicate with C/C++ programs via the Lua stack. In this method, we assumed that C/C++ would be used for the implementation, so we adopted Lua, which has a high affinity for C / C + +, as it met the requirements of this method.

6.3 Evaluation

The MEC-Executulator is the equivalent of the conventional MEC worker when executing C/C + + functions (C/C + +), and when using the MEC-Executulator, we compared the case where the load_model function is executed, which is assumed to be the first time the same process is executed (MEC-Executulator (1st)), and the case where the load_model function is not executed, which is assumed to be the second or subsequent time the same process is executed (MEC-Executulator (2nd)). In both cases, we used the average value when the process was executed 100 times. The evaluation environment was MEC Requester (Raspberry Pi 4B, ARM Cortex-A72 (4 cores, 1.5 GHz)), Server (AMD Ryzen 9 3950X (16 cores, 3.5 GHz)), and Worker (Alveo U50, Intel Core i5-12400(6 cores, 2.50 GHz)). As in the previous chapter, the MEC-Requester was assumed to be a computer mounted on a robot, and a Raspberry Pi 4B was used. The MEC-Server was equipped with sufficient memory and CPU to prevent performance from becoming a bottleneck. The MEC worker was Alveo U50. When using Alveo U50, the Vitis-AI Docker container is started on the host PC equipped with Alveo U50, and the AI application using the Vitis-AI library is compiled in that container. To run AI applications using the Vitis-AI library on the M-KUBOS board, you need to perform cross-compilation using the Petalinux tool on a computer separate from the M-KUBOS board and then transfer the executable file to the M-KUBOS board.

Figure 3.16 shows the results of offloading using the MEC-Executulator on an Alveo PC. The response time was larger for the MEC-Executulator (1st) than for C/C + +. On the other hand, MEC-Executulator (2nd) reduced the response time by 69.7% compared to C/C + +. From the results, it can be said that the overhead of MEC-Executulator does not affect performance, as the results for C/C + + and MEC-Executulator (1st) were not significantly different. Next, in MEC-Executulator (2nd), the overhead of loading the trained model was reduced because the load_model function was not executed. This reduced the response time compared to the conventional method of C/C + +.

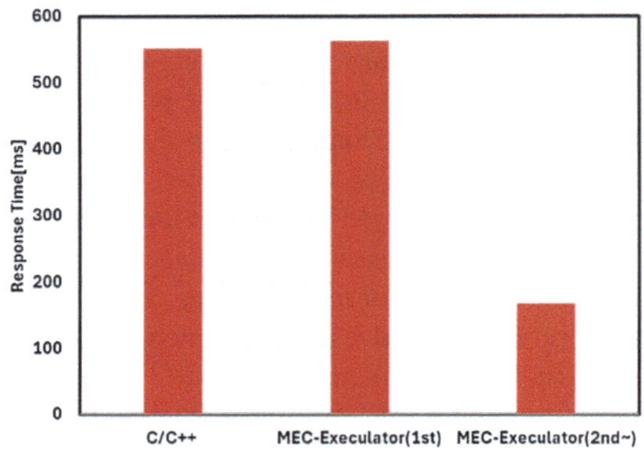

Fig. 3.16 MEC FPGA FaaS design example

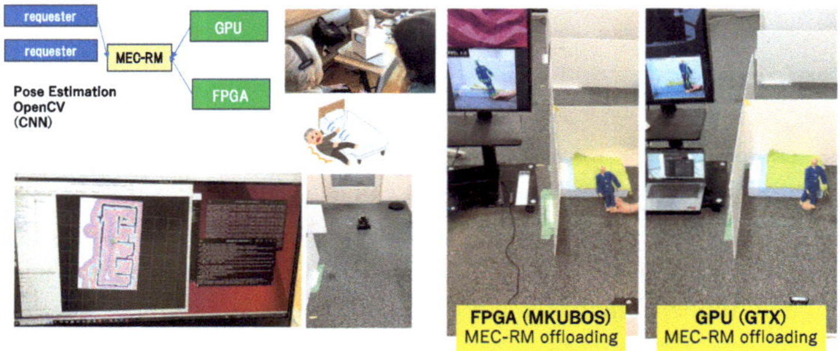

Fig. 3.17 An offloading system with MEC-RM

6.4 Developing an Integrated System

In 2024, research and development were advanced with the aim of deploying multiple robots in nursing facilities using MEC-RM [28, 29]. This integrated system uses MKUBOS, CPUs, GPGPUs, and other heterogeneous accelerators developed in this project as workers through MEC-RM, allowing the execution of posture estimation (OpenPose), which is crucial for fall detection in nursing facilities as a requester application. By detecting skeletons with OpenPose, the falls can be detected in real time. To demonstrate this, we developed a miniature model of a nursing facility. Figure 3.17 showed the small example of the proposed system. In this integrated system, MEC-RM installed on the server, robots go around the room at nursing care home, and monitoring result will be transferred to the MKUBOS (FPGA) and GPUs through MEC-RM seamlessly. If it detected the fall down of the elderly, the quick action would be taken by the robot and the caregiver.

For the posture estimation application, the processing on a standard PC with a Core i5 typically runs at about 0.5 fps. However, by offloading the processing to MEC-RM, it was confirmed that processing could be performed at 4–5 fps, making practical implementation feasible. Moving forward, it will be possible to seamlessly

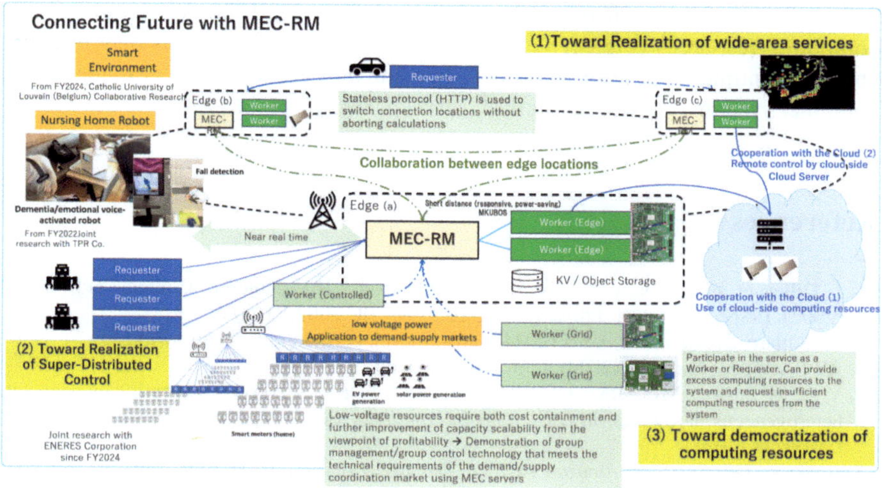

Fig. 3.18 MEC-RM world: for anybody anytime

offload high-load functions such as map generation and map matching within SLAM. These robot systems are being developed in our teams for robot systems. In addition, Nishiken's k-anonymization has been integrated into MEC-RM, and Nakadai Lab's HARK has been incorporated into robots to develop systems that approach the direction of a voice and initiate communication, promoting vertical integration. With the expectation of extensive AI processing, as shown in Fig. 3.18, MEC-RM provides a versatile interface that allows offloading functionality to be utilized by anyone, anywhere, as long as they can connect via HTTP, fostering the development of numerous systems.

7 Conclusion

In this chapter, we have examined the requirements for the necessary middleware to make MEC a truly useful service provisioning platform for society. MEC has many technical issues to address to provide high responsiveness to services for edge devices and to support services for a very large number of edge devices with high scalability. We proposed MEC-RM (MEC resource manager) to address these technical issues. MEC-RM is a general-purpose computer resource management system that can be accessed from anywhere via HTTP and that can achieve high scalability. In this section, we introduce the MEC-RM scheduling, the implementation of multiple FPGAs and FaaS, and the application of specific applications. MEC-RM can incorporate these developments and further develop in the following directions: (1) toward the realization of wide-area services, (2) toward the realization of superdistributed control, and (3) toward the democratization of computing resources (Fig. 3.18). We are already developing in this new direction and will continue to develop in order to become the cornerstone of next-generation computer systems.

References

1. J. Dilley et al., Globally distributed content delivery. IEEE Internet Comput. **6**(5), 50–58 (2002). ISSN: 1089-7801. https://doi.org/10.1109/MIC.2002.1036038
2. Cloudflare, *Cloudflare Workers* (2024). https://developers.cloudflare.com/workers/
3. AWS, *AWS for the Edge* (2024). https://aws.amazon.com/edge/
4. Y. Li, M. Sugaya, MEC-RM: a distributed job scheduler for serverless computing on edge with heterogeneous hardwares, in *IEEE 43th International Conference on Consumer Electronics*. Las Vegas (2025)
5. Amazon Web Services, *Amazon EC2 On-Demand Pricing* (2024). https://aws.amazon.com/ec2/pricing/on-demand/?nc1=h_ls
6. J. Zhang, K.B. Letaief, Mobile edge intelligence and computing for the internet of vehicles. Proc. IEEE **108**(2), 246–261 (2020)
7. A. Ndikumana et al., Deep learning based caching for self-driving cars in multi-access edge computing. IEEE Trans. Intell. Transp. Syst. **22**(5), 2862–2877 (2021). https://doi.org/10.1109/TITS.2021.3051263
8. S. Liu et al., Edge computing for autonomous driving: opportunities and challenges. Proc. IEEE **107**(8), 1697–1716 (2019)
9. AWS, *Lambda Extensions API* (2023). https://docs.aws.amazon.com/lambda/latest/dg/runtimes-extensions-api.html. Accessed 4 Apr 2025
10. S. Ristov, C. Hollaus, M. Hautz, Older than the warm start and warmer than the cold start! Experience the spawn start in FaaS providers, in *ApPLIED '22: Proceedings of the 2022 Workshop on Advanced Tools, Programming Languages, and PLatforms for Implementing and Evaluating algorithms for Distributed Systems* (2022), pp. 35–39
11. A. Kumari, B. Sahoo, R.K. Behera, Mitigating cold-start delay using warm-start containers in serverless platform, in *2022 IEEE 19th India Council International Conference (INDICON)* (2022), pp. 1–6
12. A. Fuerst, P. Sharma, Locality-aware load-balancing for serverless clusters, in *HPDC '22: The 31st International Symposium on High-Performance Parallel and Distributed Computing* (2022), pp. 227–239
13. M. Abdi et al., Palette load balancing: locality hints for serverless functions, in *EuroSys '23: Proceedings of the Eighteenth European Conference on Computer Systems* (2023), pp. 365–380
14. M.G.L. Freccassetti, R. Pagliarani et al., *Worldwide Analysis to Promote and Facilitate the Wireless Transport Network as the Key Enabler for Fast Mobile Backhaul Network Modernization*. White Paper 58. ETSI, 2023
15. Github, *AzurePublicDataset* (2023). https://github.com/Azure/AzurePublicDataset/tree/master. Accessed 4 Mar 2025
16. D. Ustiugov, T. Amariucai, B. Grot, Analyzing tail latency in server-less clouds with STeLLAR, in *IEEE International Symposium on Work-load Characterization (IISWC)* (2021), pp. 51–62
17. Labour Ministry of Health and Welfare (MHLW), *About the Population of Our Country* (2023). https://www.mhlw.go.jp/stf/newpage_21481.html?SLANG=ja&TLANG=en&XMODE=0&XCHARSET=utf-8&XJSID=0. Accessed 20 Oct 2023

18. R. Hayashi, O. Komazawa, *Health and Long-term Care Information in Ageing Asia*. Tech. rep. 07. ERIA, 2022
19. N. Bredeche, N. Fontbonne, Social learning in swarm robotics. Philos. Trans. Roy. Soc. B: Biol. Sci. 377–1843 (2021)
20. M. Dorigo, G. Theraulaz, V. Trianni, Swarm robotics: past, present, and future [point of view]. Proc. IEEE **109**(7), 1152–1165 (2021)
21. S. Macenski et al., Robot operating system 2: design, architecture, and uses in the wild. Sci. Robot. **7** (2022)
22. J. Kennedy, R. Eberhart, Particle swarm optimization. in *Proceedings of IEEE the International Conference on Neural Networks*, vol. 4. (1995), pp. 1942–1948
23. M. Quigley, B. Gerkey, W.D. Smart, ROS: an open-source robot operating system, in *Proceedings of the ICRA Workshop on Open Source Software* (2009)
24. I.T. Inage et al., M-KUBOS/PYNQ cluster for multi-access edge computing, in *CANDAR* (2021), pp. 95–101
25. AMD Xilinx, *DPU for Convolutional Neural Network* (2024). https://japan.xilinx.com/products/intellectual-property/dpu.html. Accessed 15 Dec 2024
26. AMD Xilinx, *Vitis AI* (2024). https://www.xilinx.com/products/design-tools/vitis/vitis-ai.html. Accessed 15 Dec 2024
27. H. Mori et al., Offloading image recognition processing for care robots to FPGA on multi-access edge computing, in *2023 International Conference on Field Programmable Technology (ICFPT)* (2023), pp. 270–271
28. L. Yanzhi, M. Sugaya, Proposal for a resource management system for mixed multi-FPGA and GPGPU environments. Res. Rep. High Performance Comput. **2022**(2), 1–10 (2022)
29. L. Yanzhi, M. Sugaya, Proposal and implementation of MEC-RM, a resource manager for effective use of computing resources. Res. Rep. Syst. Archit. **2024**(17), 1–9 (2024)

Chapter 4
Application for a Smart City/Community

Hiroaki Nishi and W. A. Shanaka

1 Introduction

Here, we discuss the concept of a smart community, the types of information that should be utilized within it, and the potential societal transformations it may bring. These aspects will be examined from the perspective of multi-access edge computing (MEC), incorporating real-world case studies.

In a smart community, various forms of information are collected and utilized, such as smart meters electricity consumption data, environmental and lifestyle information from home energy management systems (HEMS), user behavior data, meteorological data, and medical information. Through cross-domain analysis, new insights can be obtained and shared, enabling the development of innovative services that may even extend beyond their original intended applications.

For the successful deployment of smart community services, it is crucial to address privacy issues, establish the necessary infrastructure, and design-specific service models. MEC plays a significant role in handling these challenges. Here, we will explore the structure of the information infrastructure for smart communities, including the anticipated services and their implementation.

2 What Is a Smart Community?

First, we define the concept of a smart community. The origins of this concept can be traced back to the microgrid (μ-grid) proposed in 2002 by Chris Marney

H. Nishi (✉) · W. A. Shanaka
School of Science for Open and Environmental Systems, Center for Information and Computer Science, Faculty of Science and Technology, Keio University, Tokyo, Japan
e-mail: west@keio.jp; shanaka@west.sd.keio.ac.jp

H. Amano (ed.), *Computation Platforms for Multi-access Edge Computing*,
https://doi.org/10.1007/978-981-96-8935-4_4

at Lawrence Berkeley National Laboratory. Unlike conventional large-scale power grids, a μ-grid is a small-scale grid that incorporates distributed power sources, including renewable energy.

Since μ-grids operate with independent power generation and supply, they are self-sufficient and resilient to external disruptions, such as power outages in the main grid. However, unlike large-scale power grids, a μ-grid handles smaller volumes of electricity, making it more susceptible to fluctuations in power generation and consumption. To ensure stability, it requires highly responsive management and operation supported by Information and Communication Technology (ICT). Moreover, since the scope of the data it handles is confined within the grid, its regional specificity is high.

The concept of using ICT for power grid management is also applicable to large-scale power networks. For example, online management of distributed power sources and electricity demand can enable the construction of systems that dynamically adjust power generation, transmission, distribution, and storage while considering price fluctuations through dynamic pricing. In addition, advanced fault detection systems can be developed using network time synchronization.

In 2007, the Energy Independence and Security Act (EISA) was enacted in the United States, emphasizing key initiatives such as:

- Upgrading aging power grids
- Addressing the rising electricity demand due to population growth and increased appliance usage
- Tackling global warming issues
- Strengthening international competitiveness by modernizing power infrastructure

Subsequent large-scale funding was allocated to these efforts, accelerating initiatives to build reliable, safe, and efficient power grid infrastructure that also promotes the adoption of renewable energy. As a result, the concept of smart grids emerged, integrating ICT into power infrastructures to enhance functionality, intelligence, and integration. This movement gained traction worldwide.

In Japan, particularly after the Great East Japan Earthquake, smart grids have attracted significant attention due to the rising cost of electricity, the push for the adoption of renewable energy, and the shift toward the liberalization of the electricity market. Consequently, various smart grid initiatives were launched across the country.

The concept of smart grids, which involves the integration of ICT in the power infrastructure to improve its intelligence and efficiency, has also expanded to other domains of infrastructure [1]. For example:

- Smart transportation focuses on optimizing logistics, transportation networks, and road infrastructure through advanced traffic management systems.
- Smart agriculture employs ICT-driven field servers to improve farming efficiency.
- Smart government integrates digital solutions, such as e-governance platforms.

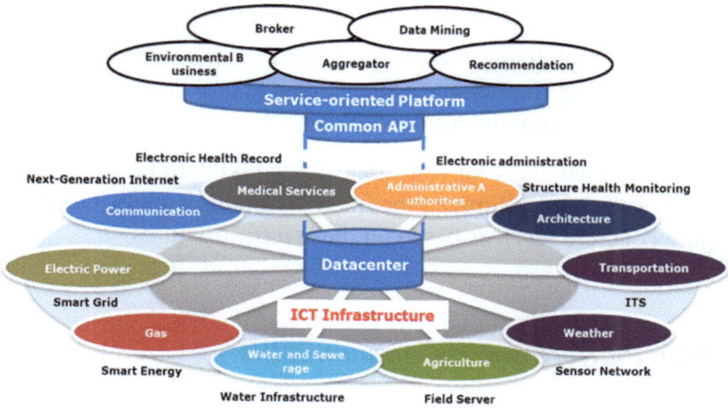

Fig. 4.1 Smart community data and services

- Smart healthcare leverages technologies like electronic medical records (EMR) to enhance medical services.

These initiatives collectively fall under the umbrella of smart infrastructure (Fig. 4.1).

When multiple smart infrastructures are concentrated in a specific region, they are referred to as a smart city, smart town, smart village, or smart island, depending on the scale of the municipality or area in which they are implemented. In contrast, the term smart community does not refer to the size of the region but instead focuses on initiatives aimed at improving local society.

In a μ-grid, discussions revolve around independence from large-scale power grids and resilience to global disruptions. A similar concept applies to smart communities, where efforts are directed toward localized information services and initiatives to improve the quality of life (QoL) of local residents.

3 The Structure of Smart Community Platforms

Traditionally, infrastructure-based services have been limited to the specific infrastructure within which they were implemented. In contrast, smart communities differ in that they integrate information from multiple infrastructures across domains. Examples of such cross-domain applications include:

- Aggregators that consolidate information
- Data mining services that transform vast amounts of unstructured data into meaningful insights
- Brokerage and recommendation services that leverage data for business applications

- Eco-business models that optimize resource consumption

A crucial factor in realizing these applications is data fusion between different infrastructures, ensuring smooth and efficient data handling.

At present, three key initiatives are necessary:

1. Developing a service platform with standardized APIs to facilitate data fusion
2. Building infrastructure optimized for smart community applications
3. Developing concrete services based on these infrastructures

3.1 *Service Platforms and Common APIs*

We first discuss (1) the development of service platforms and standardized APIs. Historically, proprietary systems have dominated the market. However, in recent years, an increasing number of open-standard systems have been proposed.

In 2006, we implemented an energy management system (EMS) that integrated small-scale local servers, low-power environmental sensor networks, and power consumption control mechanisms [2]. At the time, no established data representation standards existed. Therefore, we proposed an open sensor information format and a command expression protocol using XML [3].

This system was later extended for:

- Building EMS (BEMS) for commercial buildings [4]
- Home EMS (HEMS) for residential energy management [5]

Due to the adoption of common protocols, we could easily integrate and scale these systems into a large-scale cluster/community EMS (CEMS) [6].

Over the past few years, numerous standardization efforts have emerged, including XML-based protocols such as IEEE 1888 [7] and OpenADR [8], which align with our original design principles. As standardization progressed across various infrastructure domains, redundancies became apparent. The reasons for these redundancies include:

- Independent standardization efforts by multiple organizations addressing similar technologies
- The emergence of broader, more comprehensive standards that overlap with existing ones
- Ambiguous boundaries between layers, leading to overlapping functionalities

As a result, even though different systems adopt technical standards, interoperability issues frequently arise due to diversity in implementations.

To resolve this, a higher-level common API and a new unified platform are required to bridge the gaps between these standards. Only then can the aforementioned services and applications be effectively developed.

Efforts to standardize this concept are currently underway, with frameworks such as IEEE 1950.X, IEEE 2413, IEEE 2668, and IEEE P2805.X emerging.

Our research team actively contributes to these standardization efforts, including IEEE 1451.1.6 and IEEE P2992, while leading working groups focused on smart community components. These developments indicate that information transmission and reception mechanisms are now reaching the level of technical standardization.

3.2 Smart Community Infrastructure

We now discuss (2) the construction of infrastructure optimized for smart community applications, using distributed power control for smart grids and real-time control in autonomous driving systems as examples. In such applications, various types of real-time data must be acquired, including:

- Power system status (e.g., distributed energy sources, electricity demand)
- Automotive data (e.g., vehicle movement, road conditions, traffic status)

A traditional cloud-based system would involve collecting this data at a data center, processing it, and then issuing control commands. However, this approach may introduce latency issues [9].

For example, in Japan, a stable electricity supply requires real-time responsiveness. The Electricity Business Act mandates compliance with specific voltage and frequency error limits (e.g., 101 V ±6 V and frequency errors below 1 Hz). To meet these requirements, a sequence of processes—including error detection, notification to the data center, computation of corrective commands, transmission to power grid stabilization devices, and actual control execution—must be completed within 10 milliseconds.

IEC standards such as IEC 61000-4 require 10ms measurement intervals for power control, while IEC 61850-9-2 defines synchronization accuracy as 1 microsecond (1 μs). In Japan, there is no specific numerical regulation, but Tokyo Electric Power Company (TEPCO) enforces a strict frequency control rule of 50.0±0.2 Hz.

These examples highlight that control precision varies across applications, requiring system designers to carefully tailor their implementations based on use-case requirements.

Regardless, such low-latency processing is infeasible in cloud-based systems. Similar latency constraints apply to autonomous driving systems and industrial automation. Key concepts in power system stability Two key concepts in power grid management are:

1. Grid Stability—Maintaining stable voltage and frequency despite demand fluctuations and variable renewable energy generation
2. Supply-Demand Balance (Simultaneous Balancing)—Ensuring that total electricity supply matches demand within a 30-minute window (primarily for pricing and trading purposes)

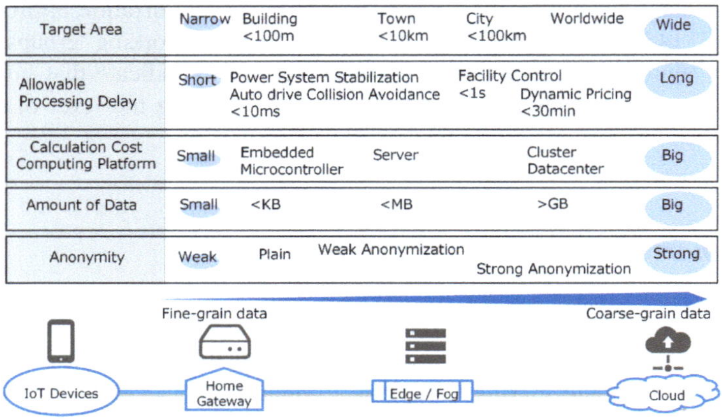

Fig. 4.2 Correspondence between information granularity and applicable scope

In large-scale power grids, simultaneous balancing is crucial. In small-scale power grids, grid stability is the priority. Since different applications operate on different time scales, system designers must consider the appropriate granularity for data collection and processing.

Application of Granularity in Smart Community Services Apart from latency constraints, other factors such as computational cost, anonymization level, and data volume also vary depending on data granularity, as illustrated in Fig. 4.2.

To optimize smart community systems, service designers must align their implementations with application-specific requirements, ensuring that the chosen data granularity matches the service objectives.

Improper granularity selection can lead to unnecessary costs in system implementation. Figure 4.3 provides a comprehensive view of major smart community services and their corresponding scope of applicability. The column denotes the items of the index. The raw denotes the granularity of target indexes. In general:

- Services applicable to larger areas (right side of Fig. 4.3) tend to tolerate higher latency.
- Services requiring lower latency (left side of Fig. 4.3) often involve higher computational costs and larger data volumes.

This analysis underscores the importance of choosing the right granularity for each service type in smart communities.

Similarly, the broader the scope of a service, the less impact increased anonymity has on the service itself. For instance, in a smart grid, obtaining detailed information such as the usage status of household appliances in each home and managing all such data at a cloud server positioned at the upper level of the hierarchy is impractical. This type of information contains detailed aspects of individuals' daily lives [10], and it has been suggested that attributes such as gender, age, and income could be inferred from it. Consequently, many households may be reluctant to

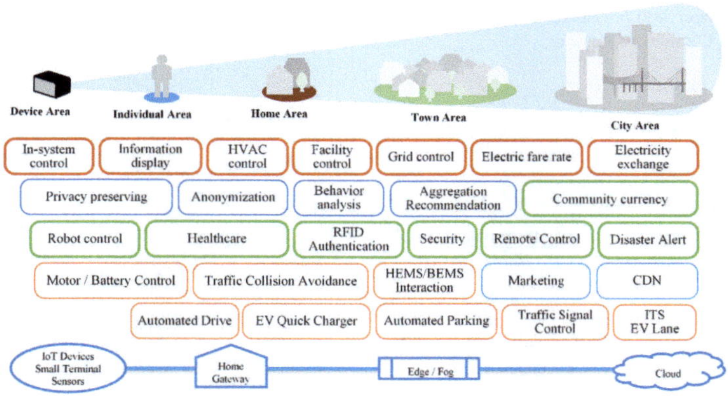

Fig. 4.3 Mapping example of smart community services

provide such data. Moreover, in achieving supply-demand balance in power grids, detailed electricity consumption data at the household level may not be necessary. In power distribution networks, detailed information may be required for small local areas managed by 6.6 kV high-voltage distribution substations. However, at 33 kV secondary substations, which manage a broader area at a higher voltage level, such detailed data is not essential.

On the other hand, there are services that do require access to detailed information, such as monitoring services and recommendation-based energy management systems (HEMS Concierge) [11], which encourage energy-saving behaviors within households. At the household level, i.e., the lowest level of the hierarchy, original non-anonymized data should be managed as locally as possible. At higher levels of the hierarchy, data management should focus on anonymizing information for applications that do not require specific details but instead handle large-scale and wide-ranging data [12]. Similarly, in autonomous driving systems, local processing may require detailed information such as vehicle positions, driving behavior, traffic signal statuses, and precise road conditions. However, at a broader scale, overall traffic congestion levels and general traffic flow patterns become more relevant than individual details.

Here, the distinction between anonymization and concealment (obfuscation) should be explained. These terms are often confused, and arguments suggesting that anonymization becomes unnecessary if data is concealed can be observed. However, the two concepts are fundamentally different. Concealment refers to techniques that keep information hidden to prevent data leakage. In contrast, the fundamental principle of anonymization is information disclosure—it is a method that allows data to be broadly utilized while ensuring that personally identifiable information is not exposed. If raw data were disclosed without processing, personal information would be at risk of leakage. Therefore, anonymization techniques selectively modify personal information or similar attributes to ensure that individuals cannot be

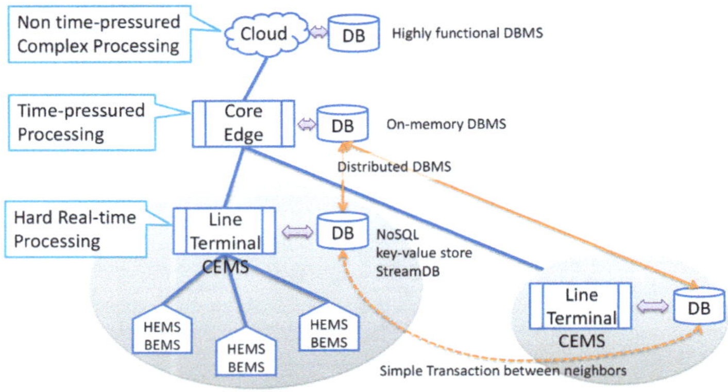

Fig. 4.4 Layered smart community infrastructure

identified, thus rendering the data safe for public release. Research and practical implementations of such anonymization techniques are extensive, including differential privacy, k-anonymity, l-diversity, and m-invariance. It is crucial to recognize that perfect anonymization does not exist, and there is a trade-off between the degree of data modification and the level of privacy protection. The more the data is altered, the lower the quality of services that rely on that data. This trade-off necessitates a data exchange infrastructure that balances data value with service quality [13, 14].

Figure 4.4 illustrates a system designed to address this consideration, leveraging a hierarchical network for appropriate processing not only at cloud servers and computational servers but also through dedicated intermediate servers within the network. This includes routers capable of directly handling application data and converting it into general services [15, 16]. When services are provided by hosts located between end-user devices and the cloud, those hosts are referred to as edge or fog nodes, and the processing performed there is known as edge computing or fog computing. In smart communities, determining the optimal location for computational processing is critical, and communication infrastructure must support multiple processing locations to maintain flexibility in information handling [17].

Thus, when designing services, it is essential to consider their scope, the degree of private information handled, and the computational burden involved. From this perspective, the concept of anonymization in smart grids and its standardized format were proposed and submitted as deliverables to the ITU-T FG-SSC [18]. Similar proposals have also been made in the industrial sector [19, 20].

3.3 *Utilization of MEC*

In Fig. 4.4, multi-access edge computing (MEC) is considered effective for locations requiring hard real-time processing or for low-latency processing in core-edge networks. A commonly raised question in discussions is whether MEC and FPGA can compete with high-performance CPUs and GPUs. While FPGAs can achieve competitive power efficiency, they often struggle to surpass CPUs and GPUs in terms of computational throughput. However, in scenarios involving communication processing, FPGAs offer clear cost advantages. Since FPGAs can be directly connected to MAC and PHY controller devices and handle communication data streams directly, they enable high-efficiency network processing and low-latency operations by supporting parallel processing of multiple communication packets. This is significantly more lightweight compared to conventional methods, where data is transferred between the main CPU and other components via internal buses or bridge chips and then processed within the OS kernel through scheduling. This distinction underscores the importance of discussing optimal hardware deployment, taking into account the granularity of information and its applicable scope, as mentioned in Fig. 4.3.

Two concrete examples illustrate this point. The first is the MQTT broker hardware. MQTT is a widely adopted lightweight protocol in the IoT domain [21]. However, since the broker is responsible for managing all message distribution, its processing load has a substantial impact on overall system performance. To address this, an MQTT hardware broker accelerator was developed for edge environments. Built using FPGA technology, this accelerator reduced latency to approximately 1/200 of that of a software-based MQTT broker, significantly enhancing performance, particularly under high-traffic conditions [22]. Additionally, a method has been established to strategically deploy these hardware MQTT brokers along communication pathways to the cloud, thereby constructing a more efficient MQTT network infrastructure [23]. Second, while anonymization is an increasingly critical technology, reducing its computational cost remains a significant challenge. Consequently, dedicated accelerators for anonymization processing have been proposed [**25**, 24].

To facilitate service provision at the edge, IoT devices and other user terminals at the network periphery must be capable of directing their traffic to edge nodes rather than traditional cloud servers. As previously discussed, the optimal processing location may shift dynamically based on optimization results, requiring flexible redirection mechanisms. Software-defined networking (SDN) is one possible approach. However, imposing high-overhead protocols on edge IoT nodes, requiring additional software installations, or replacing existing devices could introduce prohibitive costs due to the vast number of IoT devices and their typically low-cost nature. Therefore, a system must be devised that allows edge selection without requiring modifications to existing IoT devices. To address this issue, a fundamental technology has been established for migrating containerized applications while maintaining complete network consistency—ensuring that no contradictions arise in network operations before and after migration [25, 26]. Furthermore, a method has been developed that incorporates load balancing while dynamically assigning applications to the appropriate edge nodes during migration [27]. This system allows edge nodes to inspect packet contents and process relevant data in place of cloud servers. Additionally, it supports dynamic relocation of

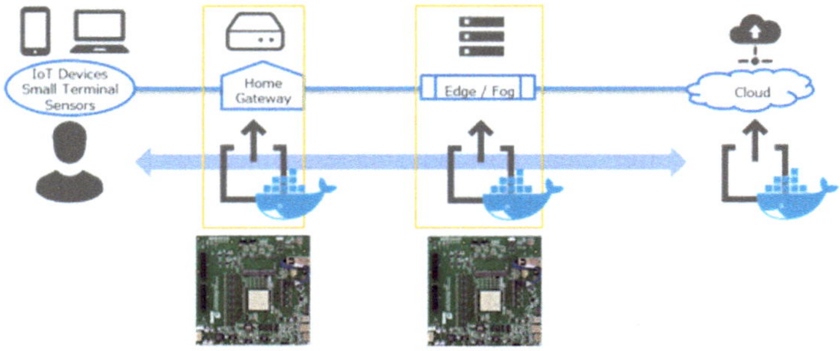

Fig. 4.5 Edge computing using MEC or containers

processing locations while ensuring network consistency is not compromised. In use cases requiring advanced acceleration, MEC is expected to play a significant role. Moreover, container-based solutions can provide flexibility to accommodate various applications. Given both considerations, the use of FPGAs emerges as a natural choice. If a system is developed that enables the seamless coexistence and integration of these technologies, it could serve as an optimal solution (Fig. 4.5).

3.4 Services Utilizing Edge Computing and Their Benefits

The specific advantages of edge computing from the service provider's perspective can be illustrated using a household electricity consumption recommendation service as an example. The envisioned service aggregates electricity usage data collected via smart meters and automatically generates electricity usage reports that highlight differences in consumption compared to other households and suggest potential energy-saving measures.

Conventionally, such services aggregate electricity consumption data in the cloud, where electricity usage reports are generated. However, in this approach, nonintrusive load monitoring (NILM) can extract information about the usage of individual appliances, and even more private details, such as household composition, could potentially be inferred. While encryption and other concealment techniques can protect data confidentiality, this assumes that the cloud service is fully trusted, as the unprocessed data is still accessible in the cloud. Given past incidents involving data breaches, it cannot be assumed that all users will be convinced of the security of cloud-based solutions.

Anonymization at the edge is an effective alternative. There exists a trade-off between the degree of anonymization, security, and service quality. Since anonymization processes may be difficult for end users to perform independently, it becomes necessary to execute such processes at the most suitable location within the system architecture, as illustrated in Fig. 4.6.

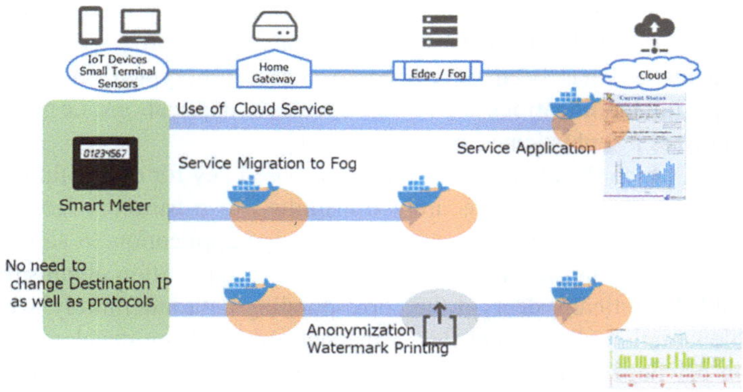

Fig. 4.6 Typical service provisioning style using edge-computing nodes

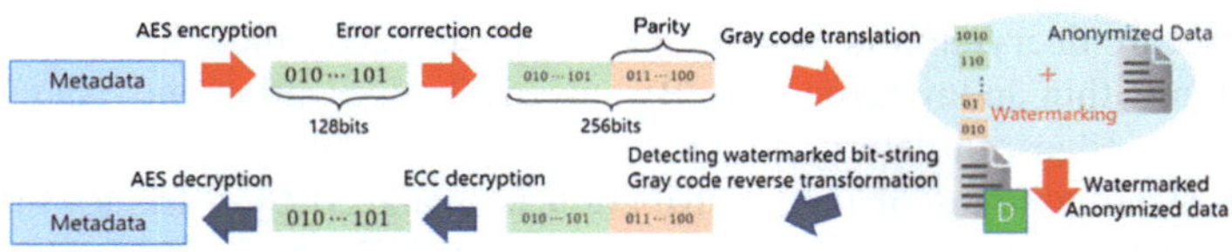

Fig. 4.7 Watermark system for data anonymization

As previously explained, the key points can be summarized into the following four aspects:

1. The first aspect is that it helps avoid an increased processing burden on users.
2. The second is that it eliminates the need for any modifications on the user side.
3. The third aspect is that, when anonymization is applied, performing the anonymization at a location where data can be sufficiently aggregated allows for a reduction in the degree of data modification, which is evaluated using a metric known as information loss. In practice, achieving these benefits requires various additional considerations, which will be discussed later.

Here, we will focus particularly on the fourth aspect. This aspect pertains to the ability of anonymization to support opt-out mechanisms required by regulations such as the GDPR, as well as to embed watermark-like distribution condition information within the data. This ensures that, even in the event of an anonymized data leak, the data's intended purpose, authorized users, and distribution scope can be determined in accordance with personal data protection laws. This is made possible by leveraging the diversity inherent in anonymization techniques [28, 29].

Figure 4.7 provides a simplified illustration of this process. If necessary, encryption can be removed, and even if the data undergoes modification, error correction codes such as turbo codes can be included to ensure the retrievability of distribution condition information. These reference values for distribution conditions are embedded using the diversity of anonymization techniques [30].

Since TLS and other encryption protocols are commonly used in actual communications, encrypted information cannot typically be processed at the edge. However, a method has been developed to decrypt encrypted data at the edge without requiring modifications to end-user devices or applications, i.e., clients. This approach utilizes a sophisticated mechanism [31].

The cloud, which the client trusts, retains a shared key for decrypting encrypted data. This shared key is then securely transmitted to a trusted edge application through a separate TLS session, enabling the edge application to safely decrypt only the necessary data [32]. However, once an edge application obtains the shared key, it could theoretically decrypt the corresponding data indefinitely, which is an undesirable situation. To prevent this, a mechanism has been developed to revoke decryption privileges for an edge application that has obtained the shared key—effectively invalidating the key it possesses. This revocation mechanism employs double encryption, ensuring that even an acquired shared key can be rendered unusable [33].

Once secure decryption is enabled, the edge can verify the embedded identifier information within anonymized data. By selectively decrypting and examining this identifier, the edge can determine whether the data is authorized for distribution. The authorization status of data distribution is managed by a framework known as vendor and consumer relationship management (VCRM). VCRM intermediates between service providers (vendors) and service consumers (consumers), managing which data a consumer provides, under what conditions, and to whom, as well as which services a vendor provides and to whom [34].

VCRM serves as an integral part of service and data provision agreements. Moreover, it advances the concept by defining which anonymized data formats qualify for specific service quality levels and how usage and data provision agreements are managed. Within the VCRM framework, users can issue opt-out requests to revoke data provision or request data deletion. While it is difficult to delete data that has already been distributed, the embedded identifier allows for restricting its further dissemination [35, 36].

Additionally, VCRM provides valuable insight into determining optimal service provision locations. It has been leveraged to propose strategies for selecting and migrating service locations when modifications are necessary [33]. Presently, FIWARE is commonly used as an integrated software platform for smart communities. Even when FIWARE is employed, a VCRM-based mechanism for managing personal data access has been successfully implemented and operationalized [37].

Furthermore, a mechanism has been established to migrate FPGA logic and internal states while maintaining network consistency in MEC environments [38]. As part of this development, a proposal has been made to migrate hardware-implemented anonymization accelerators to optimize anonymization processes dynamically and efficiently based on real-time demands [39].

4 Technical Challenges for Smart Community Infrastructure

4.1 Network Transparency and MEC Data Encapsulation

In smart cities, the concept of network transparency has been introduced to increase the potential locations for service provision, enabling services to be delivered freely at any point in the network based on demand. This concept allows for the exploration of the benefits of expanding service provision locations and ensures that terminals can maintain standard IP communication with the cloud without being aware of the specific service location. To provide edge services with network transparency and achieve low-latency processing, XDP and hardware accelerators have been enabled. Additionally, a system called "Transparent Add-on" has been developed to enhance service processing across various devices and ensure greater security. This transparent add-on makes it possible to build new smart city services that were previously unattainable. This research focused on three specific areas within this framework.

One aspect is information encapsulation, which involves anonymizing personal information at the edge before it leaves a device, location, or region. This process allows personal information to be confined to a specific area.

Another aspect is information coloring, which refers to embedding digital watermarks into anonymized information to assign it a distinct "color." This function enables, for example, the deletion of specific information at the edge based on the digital watermark, supporting the right to be forgotten.

The third aspect is transparent load balancing. With the transparent add-on, services can be provided anywhere, offering location flexibility. Additionally, 1-to-n migration becomes possible, achieving load balancing transparently. However, this requires additional measures, such as avoiding duplicate processing.

These concepts have been consolidated into a comprehensive model for an ideal smart city infrastructure. The following sections summarize the infrastructure from the cloud to IoT in a hierarchical manner.

4.2 Protection of Cloud Service and Improvement of the Design Efficiency

In this domain, there is a lack of service design methodologies that include anonymization. Specifically, designing services with anonymization requires determining where and to what extent anonymization should be applied, as these decisions significantly impact data operation policies and service quality. Therefore, a configuration that allows even nonexpert operators to design services is desirable. An effective approach involves a flexible, rapid, and modularized design methodology. To address this, a simplified, low-code system was developed. Additionally, since anonymization can degrade service quality, there is a need for tools that can

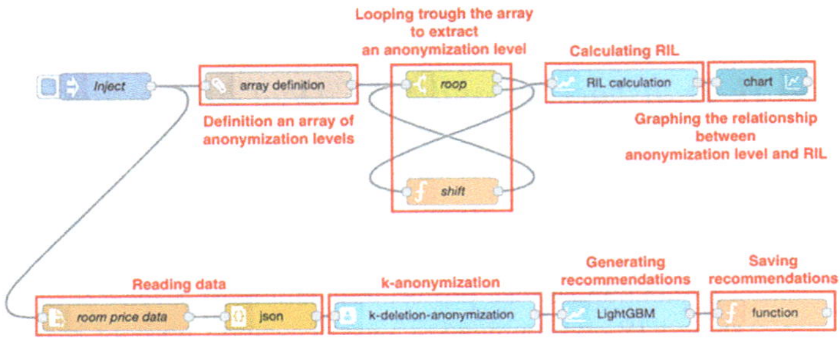

Fig. 4.8 Anonymization service tool

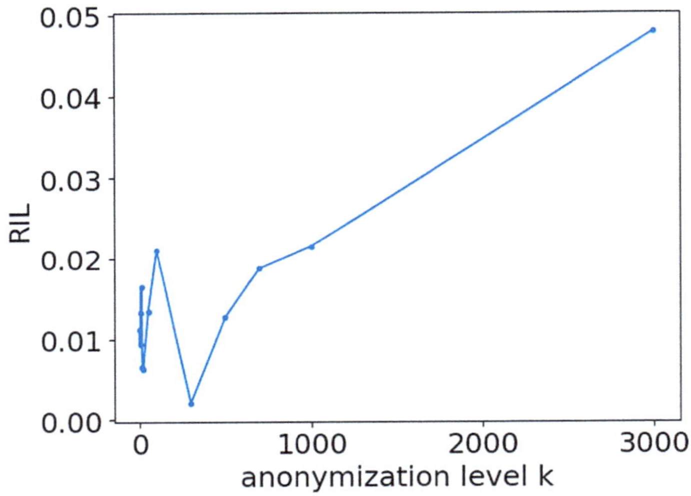

Fig. 4.9 Service quality (RIL)

be used to balance safety and service quality. Consequently, new tools and metrics were developed to assess this balance (Figs. 4.8 and 4.9).

4.3 X-as-a-Service and Their Acceleration at the Edge

The development and utilization of a smart city OS integrated with the cloud are progressing, emphasizing the need for information coloring and encapsulation. To address this, a system for monitoring the flow of personal information, integrated with edge flow monitoring capabilities, was developed. Additionally, to avoid load concentration, load balancing of processes on edge nodes was implemented. A

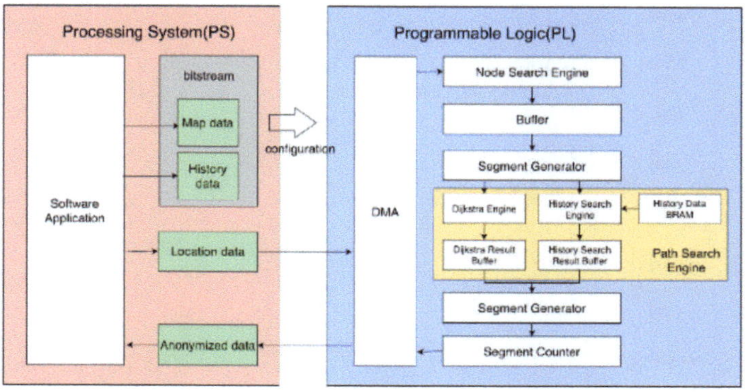

Fig. 4.10 The MQTT accelerator

system was established to migrate hardware states and logic while maintaining network consistency.

In IoT, MQTT is often used as a means of information distribution and can be considered a central point for information aggregation. Enhancing the MQTT broker to function as a core for encapsulation and anonymization is advantageous for building sensor network services. However, its low performance can hinder services requiring real-time capabilities. To overcome this, a hardware accelerator was developed, achieving approximately 111 times the throughput compared to a software-based MQTT (Mosquitto), and supporting additional anonymization features. It achieved a processing latency of 829.2 ns, including communication interfaces and MAC, and a sub-300 μs logical delay, enabling 10 Gbps line rate processing (Fig. 4.10).

Moreover, this advanced MQTT broker was made capable of hardware migration and extended into a distributed MQTT infrastructure for widespread use. This high performance enabled time synchronization using MQTT, with this approach being formalized in the IEEE P1451.1.6 standard. In practice, a time synchronization accuracy of within 52 μs was achieved among 100 subscribers (Fig. 4.11).

4.4 *Acceleration of Personal Information Protection*

As previously mentioned, anonymization on the MQTT broker was enabled, but in practical use, the data changes constantly. While the original data continues to change, anonymization requires using data at a specific point in time, which made it difficult to keep up with the changes in the original data, thus limiting the service's applicability. Therefore, we developed a new anonymization algorithm that supports τ-safety and accommodates data addition and deletion.

Based on this concept, we newly developed a system for anonymizing smartphone trajectories. Furthermore, since the software-based anonymization process

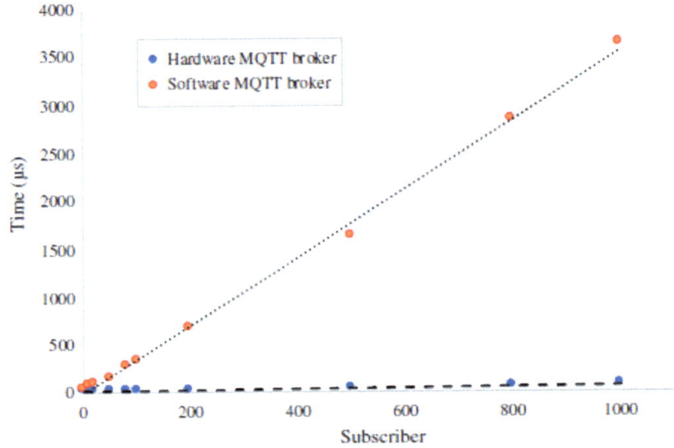

Fig. 4.11 Delay vs. the number of subscribers

takes several minutes, we built a hardware accelerator capable of completing the task in a few seconds. In this case, as well, we achieved an approximately 100-fold increase in throughput.

4.5 Processing Encrypted Information at the Edge

We enabled transparent add-on processing even when communication is encrypted. For instance, when a terminal and the cloud communicate under encrypted conditions, the edge cannot perform transparent add-on processing using the encrypted communication. To address this, the edge obtains approval from the cloud to access the common encryption key used by the terminal and the cloud. This allows the edge to join the encrypted communication and perform the add-on processing.

However, this introduces a new challenge. Once the edge gains access to this encrypted communication, it can intercept all relevant communications between the terminal and the cloud. This creates a situation where the edge cannot disengage from accessing the communication without modifications to the terminal. To resolve this, we proposed and implemented a method where adjacent edges or the cloud, in collaboration, establishes an additional layer of encrypted communication. This double encryption conceals the communication from the involved edge. After updating the secret key for the relevant communication, the double encryption is removed, effectively securing the communication from the edge (Fig. 4.12).

The transparent add-on currently supports the latest TLS 1.3, ensuring robust security. Additionally, by leveraging DPDK, we achieved low-latency processing.

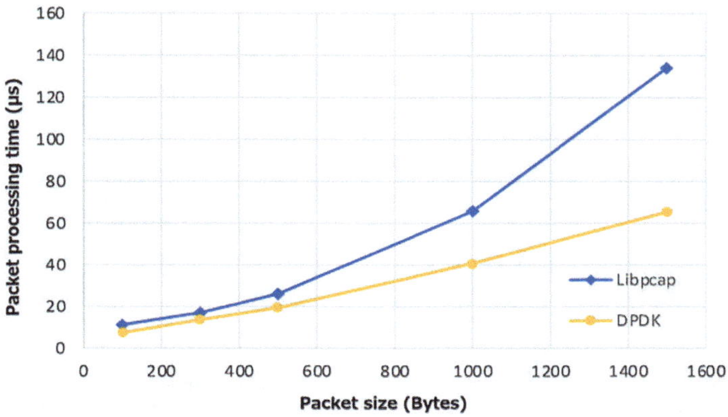

Fig. 4.12 Packet processing time vs. packet size

4.6 Dynamic Data Anonymization

In information anonymization, there have been many proposals for anonymizing fixed data. However, since data is constantly changing, there has been a need for proposals for information anonymization methods that keep up with data changes. For example, differential privacy guarantees that there will be no information leakage, no matter what kind of data change occurs, but conversely, in order to guarantee this, the data is generalized to a great extent, and the original form is almost completely lost, and the data becomes significantly less useful. k-anonymization, l-diversity, and m-eligibility, which are based on k-anonymization, do not lose much of the degree of generalization. However, they assume that the data is fixed. Therefore, in order to deal with homomorphic data, we proposed a k-anonymization method that supports the addition and deletion of data.

5 Smart Community Services

Various smart community services utilizing edge computing have been proposed. One notable example is the smart town project conducted in Urawa-Misono, a district approximately 30 minutes north of Tokyo by train (Fig. 4.13).

In this town, various types of information are collected from smart homes, including electricity consumption, environmental conditions, and lifestyle data. Since the project's launch in 2015, additional data has been integrated from commercial facilities, such as purchasing information, and from public facilities, including health metrics obtained via body composition scales. Moreover, data has been gathered using BLE-based rental bicycles and location-tracking systems (Fig. 4.14), facilitating the development of higher-level services. A critical aspect of these services is data anonymization and its integration into service frameworks.

Fig. 4.13 Smart town services in Urawa-Misono District of Saitama City

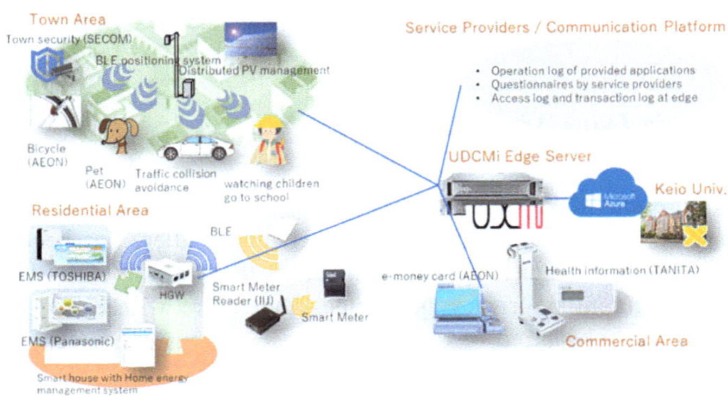

Fig. 4.14 Smart town infrastructures in Urawa-Misono District of Saitama City

On top of this environment, the previously discussed infrastructure was constructed. In this case, for simplicity, the edge hierarchy consists of only a single regional edge. However, a system was developed to facilitate the matching of various data providers and service providers, and operational testing of the system was conducted in a real-world environment (Fig. 4.15) [40].

5.1 *Constructed Infrastructure Management System*

This section provides an overview of the underlying mechanisms supporting the constructed infrastructure (Fig. 4.11). The system allows users and service providers to input the necessary information into the vendor and consumer relationship management (VCRM) framework via a web interface. Specifically, users can specify:

- What type of information do they wish to provide?

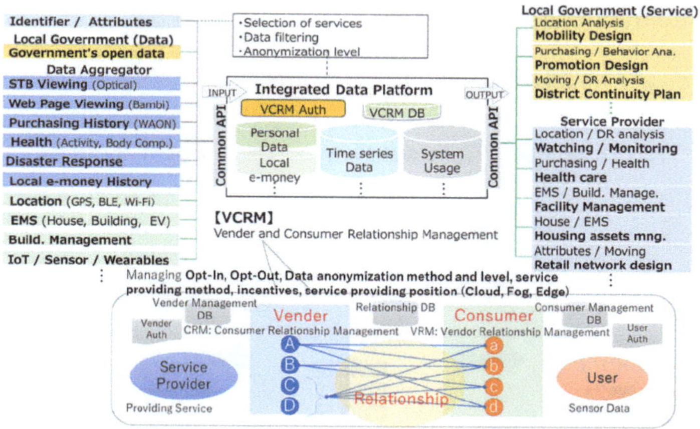

Fig. 4.15 Total diagram of data transaction and systems

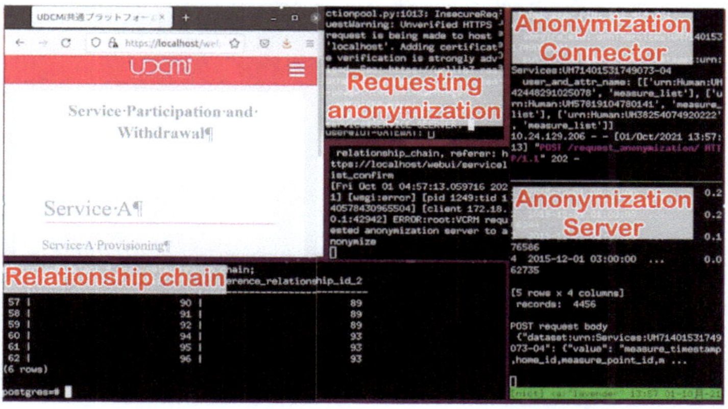

Fig. 4.16 Operation screen images

- In what format should the information be shared?
- To whom the information should be provided?
- What services they would like to receive in return?

On the other hand, service providers can define:

- What data will they use?
- How they will provide their services?

Additionally, the system enables the configuration of anonymization levels. Based on these settings, a matching process determines (Fig. 4.16):

- The cost associated with receiving services
- The compensation for providing data

Fig. 4.17 Dataflow controller

In accordance with the vendor and consumer relationship management (VCRM) framework, the data flow path is determined based on the network topology of the infrastructure. This path is maintained in the form of a relationship chain, which records the endpoints and edge nodes through which the data traverses, along with the characteristics (ID) associated with the data.

Through this mechanism, the data flow can be effectively controlled, and modifications such as data deletion or migration can be managed via a web interface (Fig. 4.17). Deletion involves obstructing the further distribution of the specified data, while modification corresponds to data migration.

In practice, since only a single network layer has been implemented, an emulation environment was also constructed to virtually increase the number of layers (Fig. 4.18).

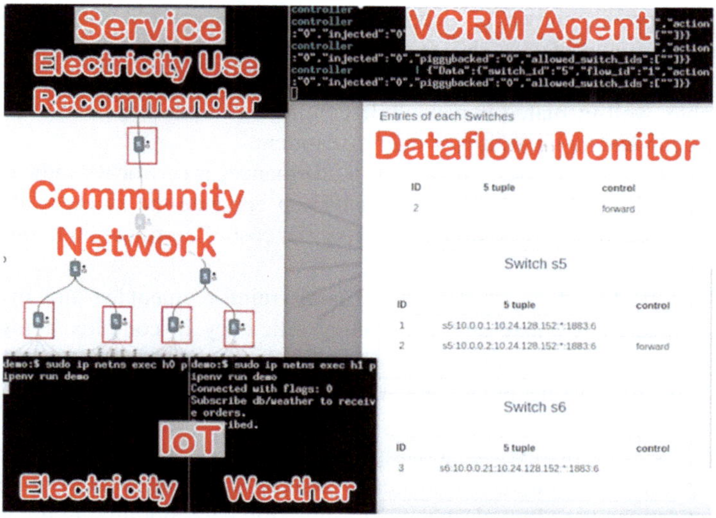

Fig. 4.18 Smart community infrastructure emulator

Within this environment, operational testing of the vendor and consumer relationship management (VCRM) system was conducted using a distributed architecture called VCRM agents to prevent excessive load concentration on the VCRM.

Furthermore, designing applications that anonymize data while providing services requires specialized knowledge of anonymization techniques. Additionally, understanding the trade-off between the degree of anonymization and the resulting degradation in service quality is crucial. A framework that facilitates the implementation of service applications with built-in anonymization, as well as a mechanism for investigating the trade-off between anonymization levels and service quality, has been proposed and implemented [41].

5.2 Examples of Constructed Services

5.2.1 Smart Town in Urawa-Misono, Saitama City

We describe experimental services developed using data from the Urawa-Misono district in Saitama City. Each of these services is capable of operating within an anonymized environment. The first service involves an evaluation system for indoor thermal environments, based on data collected from households, including room temperatures and air conditioner power consumption. This system estimates and assesses the thermal insulation performance of residences [42]. Using the estimated thermal transmittance of a given residence, homes with poor insulation performance can receive financial investment from banks. These funds are then used to carry out insulation improvement work, such as installing double-glazed windows or adding thermal insulation materials. Since this results in reduced energy costs, the banks recover their investment through the savings. Residents benefit by living in more comfortable, well-insulated homes while continuing to pay the same energy costs, and banks can ensure the return of their investment.

For homes with excellent insulation performance, a certificate indicating high insulation quality is issued. This certificate can enhance the market value of the property when selling a used home, allowing the cost of certification to be offset by the increased selling price.

Additionally, based on actual collected data, a reinforcement learning mechanism was developed to optimize air conditioning efficiency in complex living spaces, ensuring uniform temperature distribution [43]. Similarly, a short-term power consumption forecasting model was constructed using deep learning techniques [44].

Another service involves generating personalized electricity usage recommendation reports. This system automatically determines whether users followed the recommendations based on electricity consumption data, optimizing the generated reports accordingly [45].

In the district's commercial facilities, consumer purchasing data was collected and provided to companies that required such information. As compensation for data provision, residents received various services. Since the purchase data includes information on food items, the service analyzed the types and quantities of purchased food based on item names and pricing. By incorporating pre-obtained household composition data, the system assessed whether the purchased foods provided an adequate nutritional balance for the family (Fig. 4.19) [46].

This service calculates the required daily intake of nutrients, compares it to the nutrients obtained from purchased food items, and suggests additional foods needed

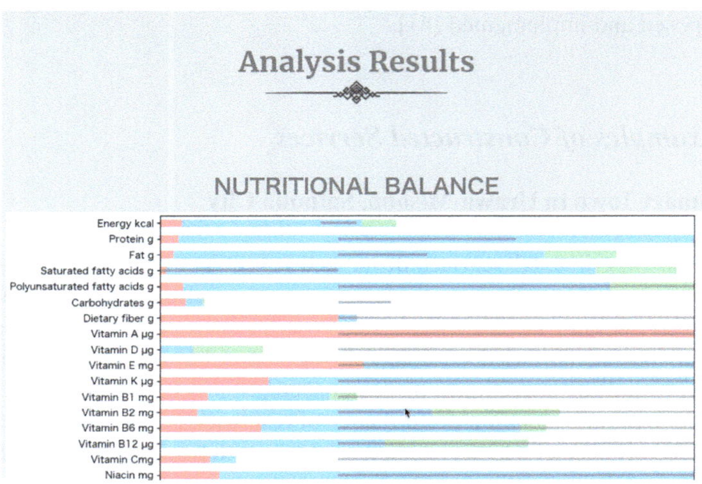

Fig. 4.19 Recommendation service of food purchasing

to meet dietary requirements. Additionally, it provides recommended recipes that optimize nutrition intake based on available ingredients.

In this district, location information was obtained using BLE technology. Unlike conventional data anonymization, location data anonymization presents unique challenges since timestamps and coordinate data form continuous trajectories. To address this, various anonymization techniques were developed. Initially, a simple noise-based anonymization method was proposed [47], followed by a deep-learning-based anonymization approach [48]. Furthermore, a hardware accelerator for anonymizing trajectories using k-anonymity was developed [49]. This accelerator successfully achieved a processing speed approximately 100 times faster than software-based approaches.

A similar framework was also implemented in rural areas. In smart agriculture, the acquisition and management of farming data are essential. It is desirable to manage this data using the same infrastructure. Various agricultural data utilization services were proposed, including the development of models for greenhouse temperature estimation [50], techniques for temperature control to achieve uniform greenhouse temperature distribution [51], and the establishment of data representation standards for service proposals based on the proposed technology [52].

5.2.2 Green Society ICT Life Infrastructure in Kurihara City, Miyagi Prefecture

Finally, we discuss the Green Society ICT Life Infrastructure initiative in Kurihara City, Miyagi Prefecture. This project goes beyond enhancing individual infrastructure sectors such as healthcare, energy, agriculture, meteorology, local government, and disaster prevention. Instead, it aims to integrate these domains into a unified societal infrastructure and explore its applications. At the core of this initiative is an information management system that collects and integrates data from various infrastructures. This system adheres to established technology standards as much as possible and provides an open interface that allows flexible data analysis and anonymization, thereby supporting the development of diverse new services.

For example, by analyzing the correlation between heatstroke incidence and meteorological data, the system can learn from past trends. Incorporating weather forecasts further enables predicting the number of future heatstroke cases in the region. This prediction can be used to estimate ambulance dispatch numbers and hospital bed requirements, contributing to more effective medical service planning.

Relatedly, heatstroke is particularly prevalent among the elderly who also exhibit high rates of recurrent hospital visits. This is often attributed to two factors: Elderly individuals either tolerate heat without recognizing its severity or experience a decline in their ability to perceive heat as they age. By integrating temperature sensor data from households and agricultural fields with medical records, physicians can gain insights into patients' past environmental conditions. This data can then be used for preventive healthcare guidance, potentially reducing hospital readmission rates.

5.2.3 Applications in Energy and Agriculture

Energy-related applications have also been proposed. For instance, using 50-year climate projections combined with current electricity demand, it has been estimated that electricity demand will increase by 4%. Moreover, projections indicate that the number of extremely hot days will increase eight-fold, leading to a rise in electricity supply shortages. This highlights the growing importance of demand response services and the need for future peak demand mitigation strategies.

Additionally, agricultural field sensing is expected to contribute to crop selection optimization and understanding the relationship between agricultural labor and health. Social Capital and Smart Community Development Beyond infrastructure, the concept of social capital—which refers to trust, social networks, and relationships among community residents—is gaining attention. Efforts have been made to quantify social capital through surveys, resulting in the creation of a social capital index.

Previous studies have established a strong correlation between healthcare and social capital indices. Higher social capital has been linked to better health outcomes in the elderly and higher rates of recovery from medical interventions. In Kurihara City, a comprehensive survey was conducted to generate a social capital map. By integrating these data with healthcare guidance, the correlation between social capital and health outcomes was validated.

Expanding on this concept, it has been suggested that introducing energy-saving systems or implementing collective electricity procurement in high-social-capital areas may yield greater effectiveness.

Moreover, by analyzing individual comfort levels in air conditioning control, it is possible not only to optimize energy efficiency, but also to assess heatstroke risk based on how individuals perceive comfort. This could further contribute to both energy conservation and public health improvement [53].

6 Conclusion of this Section

Smart communities require cross-domain data integration, and building an infrastructure that connects these data to services is of paramount importance.

As shown in Fig. 4.20, data is collected from the community (community sensing), leading to behavioral modifications and device control. Furthermore, even the results of these interventions are collected as data, enabling continuous evaluation and improvement. Through these efforts and the establishment of data cycles, solutions can be provided for current and future social challenges. By offering services that encourage desirable behavioral changes among residents, these initiatives can lead to improved quality of life (QoL) for the community.

This is the essence of a smart community infrastructure and services. Furthermore, within this infrastructure, multi-access edge computing (MEC) plays a crucial

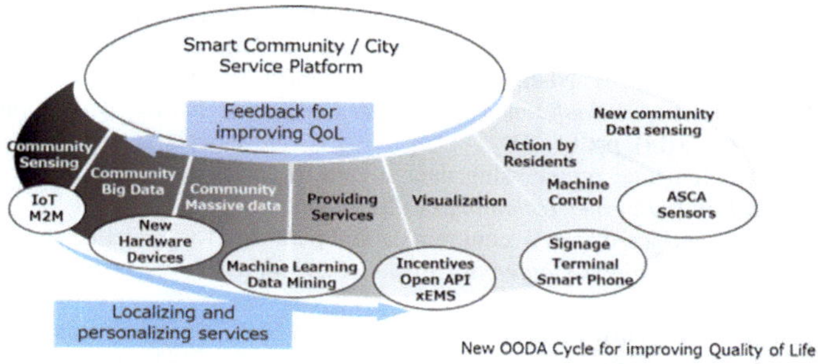

Fig. 4.20 Data cycles in smart community services

role as a solution that effectively meets the growing and specific demands of smart communities.

References

1. J. Haase et al., The IOT mediated built environment: a brief survey, in *2016 IEEE 14th International Conference on Industrial Informatics (IN-DIN)* (2016), pp. 1065–1068
2. S. Ishida et al., Implementation of Internet based Demand Controlling System, in *4th International IEEE Conference on Industrial Informatics* (2006), pp. 381–386
3. S. Ishida et al., Discussion of aspects in energy management with demand response system KNIVES, in *11th IEEE International Conference on Emerging Technologies and Factory Automation* (2006), pp. 465–468
4. T. Handa et al., KNIVES: network based demand and supply control system – enhancement for network and environmental affinity, in *5th International IEEE Conference on Industrial Informatics* (2007), pp. 1171–1176
5. Y. Suhara et al., Distributed demand side management system for home energy management, in *The 36th Annual Conference of the IEEE Industrial Electronics Society* (2010), pp. 2424–24290
6. G. Mine et al., Evaluation of cooperative demand control on micro-grid, in *The 35th Annual Conference of the IEEE Industrial Electronics Society* (2009), pp. 3629–3634
7. *IEEE STANDARD 1888-2011 - IEEE Standard for Ubiquitous Green Community Control Network Protocol* (IEEE Standards Association. 2011)
8. *OpenADR Alliance*. http://www.openadr.org/
9. K. Birman, Cloud computing for a robust, scalable, and cost-effective smart grid, in *Slides of Keynote Speech #2 in The 1st International Work-shop on Trustworthiness of Smart Grids in conjunction with the 44th Annual IEEE/IFIP International Conference on Dependable Systems and Networks* (Keynote Speech, 2014)
10. G. Wood, M. Newborough, Dynamic energy-consumption indicators for domestic appliances: environment, behaviour and design. Energy Build. **35**(8), 821–841 (2003)
11. T. Watanabe et al., HACCS: home appliance control concierge system: extending functions on web service, in *IEEE 37th Annual Computer Software and Applications Conference* (2013), pp. 208–213
12. H. Nishi, K. Inoue et al., *IEEE Smart Gird Vision For Vehicular Technology: 2030 and Beyond*. Tech. rep. IEEE Standards Association, 2014, pp. 41–42

13. K. Okada, H. Nishi, Big data anonymization method for demand response services, in *The 2014 International Conference on Internet Computing and Big Data* (2014), pp. 76–82
14. Y. Nakamura, K. Matsui, H. Nishi, Anonymization infrastructure for secondary use of data, in *The 2014 International Conference on Internet Computing and Big Data*. Conference held July 21–24, 2014 (2014), pp. 65–71
15. K. Inoue et al., Semantic router using data stream to enrich services, in *3rd International Conference on Future Internet Technologies* (2008), pp. 20–23
16. R. Kubo et al., Service-oriented communication platform for scalable smart community applications, in *38th Annual Conference of the IEEE Industrial Electronics Society* (2012), pp. 3574–3579
17. W.A. Shanaka, P. Abeysiriwardhana, J.L. Wijekoon, H. Nishi, Smart community edge: stream processing edge computing node for smart community services. IEEJ Trans. Electron. Inform. Syst. **140**(9), 1030–1039 (2020)
18. J. Chang et al., *Anonymization Infrastructure and Open Data in Smart Sustainable Cities*. Approved Deliverable of ITU-T Focus Group on Smart Sustainable Cities. https://www.itu.int/md/T13-SG05-151012-TD-GEN-1009/en
19. W. Dai et al., Industrial edge computing: enabling embedded intelligence. IEEE Ind. Electron. Mag. **13**(4), 48–56 (2019)
20. M. Bakhtiari et al., Optimum configuration of edge computing protocols for industrial Internet-of-Thing applications, in *IECON 2022 – 48th Annual Conference of the IEEE Industrial Electronics Society* (2022), pp. 1–6
21. H. Nishi, K.B. Lee, Implementation of IEEE P1451.0 and P1451.1.6 sensor networks, in *IECON 2022 – 48th Annual Conference of the IEEE Industrial Electronics Society* (2022), pp. 1–6
22. K. Yamamoto, A. Fukuhara, H. Nishi, Hardware implementation of MQTT broker and precise time synchronization using IoT devices. IEEJ Trans. Elec. Electron. Eng. **17**, 209–221 (2022)
23. T. Toyohara, H. Nishi, Distributed MQTT brokers infrastructure with network transparent hardware broker, in *2023 Eleventh International Symposium on Computing and Networking (CANDAR)* (2023), pp. 182–188
24. J. Sawada, H. Nishi, Hardware acceleration and data-utility improvement for low-latency privacy preserving mechanism, in *The International Conference on Field Programmable Logic and Applications (FPL)* (2012), pp. 499–502
25. T. Miura et al., Novel infrastructure with common API using docker for scaling the degree of platforms for smart community services, in *2017 IEEE 15th International Conference on Industrial Informatics (INDIN)*. Emden, Germany (2017), pp. 474–479
26. W.A. Shanaka P. Abeysiriwardhana et al., Consistency guaranteed multi container migration for smart community network services. Trans. Inst. Electr. Eng. Jpn. C, A Publ. Electron. Inform. Syst. Soci. **141**(12), 1453–1461 (2021)
27. Y. Namba, R. Morishima, H. Nishi, Network-transparent load balancing and migration for edge computing, in *2024 IEEE 21st Consumer Communications & Networking Conference (CCNC)* (2024), pp. 228–233
28. Y. Ohno, T. Iwai, H. Nishi, Blind watermarking method for anonymized data, in *2017 Fifth International Symposium on Computing and Networking (CANDAR)* (2017), pp. 139–145
29. Y. Nakamura, H. Nishi, Digital watermarking for anonymized data with low information loss, in IEEE Access **9**, 130570–130585 (2021)
30. Y. Nakamura, Y. Nakatsuka, H. Nishi, Novel method to watermark anonymized data for data publishing. IEICE Trans. Inform. Syst. **E100-D.8**, 1671–1679 (2017)
31. H. Hiraga, H. Nishi, Network transparent decrypting of cryptographic stream considering service provision at the edge, in *2021 IEEE 19th International Conference on Industrial Informatics (INDIN)* (2021), pp. 1–6
32. T. Fukui, Y. Sato, H. Nishi, Network-transparent decryption method in edge area compatible with TLS 1.3, in *IECON 2024 - 50th Annual Conference of the IEEE Industrial Electronics Society* (2024), pp. 1–6

33. Y. Sato, Y. Namba, H. Nishi, Layer-7 and 5-tuple information analysis framework for providing positional flexibility in location determination for service provision, in *IECON 2023-49th Annual Conference of the IEEE Industrial Electronics Society* (2023), pp. 1–6
34. A. Niwa, H. Nishi, An information platform for smart communities realizing data usage authentication and secure data sharing, in *2017 Fifth International Symposium on Computing and Networking (CANDAR)* (2017), pp. 119–125
35. S. Shimahara, H. Nishi, Dataflow management platform for smart communities using an edge computing environment, in *IECON 2021 – 47th Annual Conference of the IEEE Industrial Electronics Society* (2021), pp. 1–6
36. S. Shimahara, M. Nohara, H. Nishi, Access control management system for edge computing environment using tag-based matching and cache injection. IEEJ Trans. Electr. Electron. Eng. **18**, 1472–1481 (2023)
37. A. Fukuhara, S. Shohata, H. Nishi, FPGA context-based live migration maintaining network consistency, in *2020 Eighth International Symposium on Computing and Networking Workshops (CANDARW)* (2020), pp. 81–86
38. Y. Namba, H. Nishi, Personal data access and distribution management extension to FIWARE, in *IECON 2022 – 48th Annual Conference of the IEEE Industrial Electronics Society*. Brussels, Belgium (2022), pp. 1–6
39. T. Toyohara, K. Yamamoto, H. Nishi, FPGA live migration of content-based anonymization mechanism with network consistency. IEEJ Trans. Electr. Electron. Eng. **19**, 507–517 (2024)
40. H. Nishi, Information and communication platform for providing smart community services: system implementation and use case in Saitama city, in *2018 IEEE International Conference on Industrial Technology (ICIT)* (2018), pp. 1375–1380
41. M. Honda, H. Nishi, Anonymity-aware framework for designing recommender systems. IEEJ Trans. Electr. Electron. Eng. **19**, 1455–1464 (2024)
42. Y. Sakuma, Y. Nakajo, H. Nishi, Building thermal performance assessments using simple sensors for the green new deal in Japan, in *2018 IEEE 27th International Symposium on Industrial Electronics (ISIE)*. Cairns, QLD, Australia (2018), pp. 691–696
43. Y. Sakuma, H. Nishi, Adaptive control method of HVAC for uniformizing comfort at japanese residential living rooms using deep reinforcement learning. IEEJ Trans. Electron. C: Inform. Syst. **141**(3), 373–382 (2021)
44. Y. Sakuma, H. Nishi, Hierarchical multiobjective distributed deep learning for residential short-term electric load forecasting. IEEE Access **10**, 69950–69962 (2022)
45. Y. Takayama, Y. Sakuma, H. Nishi, Recommendation system for energy consumption behavior change on residents' response and stress, in *IECON 2021 – 47th Annual Conference of the IEEE Industrial Electronics Society* (2021), pp. 1–6
46. M. Honda, H. Nishi, Household nutrition analysis and food recommendation using purchase history, in *2021 IEEE 30th International Symposium on Industrial Electronics (ISIE)* (2021), pp. 1–7
47. Y. Sakuma et al., Trajectory anonymization through laplace noise addition in latent space, in *2021 Ninth International Symposium on Computing and Networking (CANDAR)* (2021), pp. 65–73
48. Y. Sakuma, H. Nishi, Practical trajectory anonymization method using latent space generalization. IEEJ Trans. Electr. Electron. Eng. **20**(6), 934–942 (2025)
49. H. Nakano, K. Yamamoto, H. Nishi, Accelerator for trajectory anonymization using map matching. IEEE Access **12**, 152277–152288 (2024)
50. E. Hayashi et al., Estimation of indoor space temperature distribution using heat maps, in *IECON 2023-49th Annual Conference of the IEEE Industrial Electronics Society* (2023), pp. 1–6
51. H. Nishi et al., Temperature monitoring and airflow control system for balancing the greenhouse environment using IEEE 1451 standards, in *IECON 2023-49th Annual Conference of the IEEE Industrial Electronics Society* (2023), pp. 1–6

52. M. Kozakai, E. Hayashi, H. Nishi, Integrated information representation using IEEE P2992 and its application use cases, in *IECON 2024 - 50th Annual Conference of the IEEE Industrial Electronics Society* (2024), pp. 1–6
53. H. Imamura et al., A population-based survey of social capital and instrumental ADL among community-dwelling elderly in a rural area, Japan, in *140th Annual Meeting of American Public Health Association*. Abstract No. 264501 (2012)

Chapter 5
Robot Audition

Lin Zirui, Haris Gulzar, Kazuhiro Nakadai, and Hideharu Amano

1 Introduction

Audition is the most critical aspect of human interindividual communication [1], enabling individuals to perceive sound sources under various acoustic conditions, focus on specific sources, and extract and understand information from sound. Similarly, sound is essential for robots and smart devices, providing information for navigation and interaction within their surroundings [2–7]. To enable robots and smart devices with these auditory capabilities, robot audition technology has emerged [8], which primarily realizes Sound Source Localization (SSL), Sound Source Separation (SSS), and Automatic Speech Recognition (ASR), endowing robots and smart devices with human-like auditory capabilities—recognizing sound sources, perceiving dynamic environmental changes, receiving voice commands, and facilitating effective human-machine interaction [9–11].

Robot audition technology has a wide range of applications in different scenarios. For small-scale embedded applications, robot audition systems can serve as hands-free assistants in automobiles and mobile devices and assistive devices for individuals with hearing impairments [12]. In edge computing environments, robot

L. Zirui · K. Nakadai (✉)
Department of Systems and Control Engineering, School of Engineering, Institute of Science Tokyo, Tokyo, Japan
e-mail: linzirui@ra.sc.e.titech.ac.jp; nakadai@ra.sc.eng.isct.ac.jp

H. Gulzar
Software Innovation Center, NTT, Tokyo, Japan
e-mail: haris.gulzar@ntt.com

H. Amano
Systems Design Laboratory, Graduate School of Engineering, Bunkyo-ku, Tokyo, Japan
e-mail: hunga@dlab.t.u-tokyo.ac.jp

H. Amano (ed.), *Computation Platforms for Multi-access Edge Computing*,
https://doi.org/10.1007/978-981-96-8935-4_5

audition systems can be applied to human-robot voice control and communication with in-home robot assistants and in robot scene analysis and monitor sounds for wild animal behavior analysis [12, 13].

2 Automatic Speech Recognition

ASR usually relies on high performance of ML algorithms and has enabled numerous applications based upon speech interface in our daily life, but most of the frameworks use computationally expensive algorithms deployed on cloud servers as speech recognition engines. With the recent surge in the number of Internet of Things (IoT) devices, a robust and scalable solution for enabling (Artificial Intelligence) AI applications on IoT devices is inevitable in form of edge computing. In this chapter, we propose the application of System on Chip (SoC) powered edge computing device as accelerator for speech commands classification using Convolutional Neural Network (CNN). Different aspects affecting the CNN performance are explored and an efficient and lightweight model named as CASENet is proposed which achieves state-of-the-art performance with a significantly smaller number of parameters and operations. Efficient extraction of useful features from audio signal helped to maintain high accuracy with a 6× smaller number of parameters, making CASENet the smallest CNN in comparison to similarly performing networks. Lightweight nature of the model has led to achieve 96.45% validation accuracy with a 14× smaller number of operations which makes it ideal for low-power IoT and edge devices. A CNN accelerator is designed and deployed on FPGA part of SoC equipped edge server device. The hardware accelerator helped to improve the inference latency of speech command by a 6.7× factor as compared to standard implementation. Memory, computational cost, and latency are the most important metrics for selecting a model to deploy on edge computing devices, and CASENet along with the accelerator surpasses all of these requirements.

2.1 Background and Motivation

With the advancement in robot technology and high-performance machine learning (ML) approaches, we are rapidly bridging the gap between humans and assistive robots through various human-machine interfaces, especially face expressions, speech recognition, and hand gestures [14]. Speech interface is becoming de facto interface in Human-Robot Interaction (HRI), which finds its application for assistive robots in various indoor scenarios [15] and in industrial environments to operate coworking robots [16]. A voice command controller is proposed to maneuver Unmanned Aerial Vehicle (UAV) through a basic set of commands by using an ML-based approach [17]. A Convolutional Neural Network (CNN)-based speech interface is even being tested to operate aircrafts in a simulated environment [18].

Conventional speech recognition frameworks require speech signals to be sent to a cloud server, where large ML models process the speech signals and send back the recognition results or corresponding commands. With the increasing demand for speech-enabled intelligent devices, a lot of research efforts and resources are being invested to enable high-performance speech interface on Internet of Things (IoT) devices [19]. Speech-based interface for IoT devices is also very suitable for people with mobility restrictions to operate intelligent devices at home without any movement [20].

IoT devices are usually battery-powered small devices; hence they have limited computational resources and mostly require integration with cloud engine for speech recognition tasks. Sending speech data to cloud poses a risk of privacy and data breaching. In addition to that, unstable Internet connection may cause increased latency, which is a highly undesirable scenario for practical applications of HRI through speech interface. Edge computing solves the problems by bringing the best of both worlds together at network edge. A local edge server device runs computations near the application end which makes it ideal for speech-enabled IoT devices by providing computational power at much lower latency without any risk of privacy breaching. Figure 5.1 illustrates the speech interface layout for IoT devices by making use of edge computing in comparison to using a cloud server. Shifting from centralized computing to distributed approach also helps in leveraging the fault tolerance in the system, where other edge nodes work fine when one of them faces a breakdown.

As edge computing devices are a scaled-down version of large computing clusters in cloud server, their computational power is relatively lower as compared to cloud-based computational engines. ML models have to be carefully designed and deployed on these devices to optimize their resource utilization and performance to fulfill the application requirements and resource allocation. In addition to CPUs

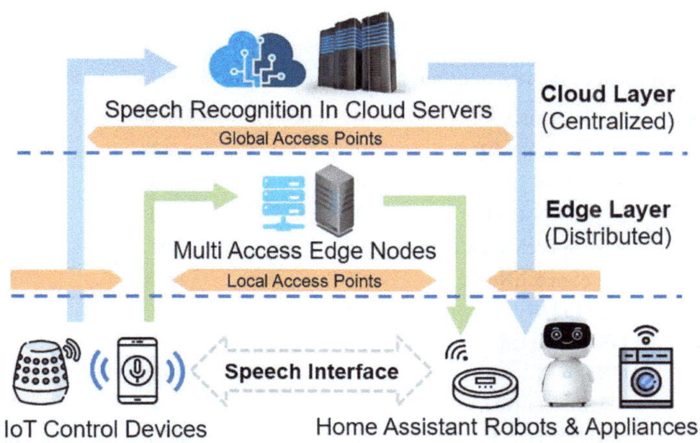

Fig. 5.1 Speech interface for IoT devices enabled by edge computing

and GPUs, ML models are also being deployed on FPGAs to meet faster inference and energy saving requirements of practical applications [21]. In addition to its flexibility to be a reprogrammable hardware, FPGA can also deliver application chip to market in short time. A System on Chip (SoC) processor is a perfect candidate for deploying ML models on edge computing devices, as it has CPU and FPGA being fabricated on the same chip. FPGA part is used as an accelerator, while CPU provides the framework for running real-time applications by making use of on-chip FPGA accelerator. Considering the application requirements of edge computing for speech-enabled intelligent IoT devices, this work makes the following contributions:

- It explores the lightest CNN architecture with the least number of parameters and operations to achieve low latency while maintaining the state-of-the-art (SOTA) performance accuracy.
- It demonstrates how efficient use of input information by feature extraction can help to achieve greater performance with lighter machine learning models.
- It introduces an accelerator for SoC-based edge computing device to achieve significantly low latency.

2.2 *Related Work*

A variety of Recurrent Neural Network (RNN) and CNN-based deep learning models have been proposed for speech commands classification on a range of CPU and GPU devices as listed in Table 5.1. A hybrid approach of RNN and attention-based approach has been adopted [22, 23], which achieves notable accuracy on edge device for speech commands classification. But the size of RNN network in terms of network parameters is too large which makes it harder to deploy these

Table 5.1 Performance metrics of different models on Google Speech-Commands Dataset V1

Model type	Model name	Accuracy (%)	Params. (K)	FLOPS (M)
RNN	MHAtt-RNN [9]	97.2	743	–
	EdgeRNN [10]	96.62	830	26.96
	EdgeRNN-G [10]	96.82	830	2.96
CNN	EdgeSpeechNetA [11]	96.8	107	343
	EdgeSpeechNetD[11]	95.8	80.3	24.5
	RES15 [14]	95.8	238	894
	DS-CNN [15]	95.4	161	56.9
	CNN [16]	96.19	1488	–
	CASENet1 (Ours)	96.45	224	**23.7**
	CASENet1-L (Ours)	95.4	**67.7**	**6.5**
	CASENet2 (Ours)	95.11	**93.2**	**3.7**

networks on resource-constrained edge devices. CNN-based approach is best suited in such scenarios as it maintains the high performance with far less number of parameters. EdgeSpeechNet [24] achieves almost the same performance as RNN with significantly fewer parameters, but the number of a floating-point operations per second (FLOPS) is increased. Similarly a CNN-based RES15 network [25] achieves comparable performance, but a number of operations increase even further.

A depth-wise separable CNN (DS-CNN) has been proposed [26] which elevates the performance of CNN by exploiting depth-wise convolutions and achieves the maximum accuracy of 95.4% for speech commands classification. Increasing network size and depth is one approach to achieve high accuracy in CNN [27], but it may lead to a large number of parameters which can easily exceed resource limitations of edge devices.

By efficiently extracting useful information from speech input, a network can achieve high performance with reduced computational effort and memory utilization. We have proposed a CNN-based network which achieves SOTA accuracy with significantly reduced network size and computational cost. A CNN accelerator is also designed and deployed on SoC-based device to achieve low latency for practical use cases. Input features extraction from speech commands dataset and introduction to our network are explained in the following sections.

2.3 Dataset and Input Features

The TensorFlow team at Google has created an open-source speech commands dataset [28]. The basic version of dataset includes ten common speech commands, i.e., Yes, No, Up, Down, Left, Right, On, Off, Stop, and Go. This dataset serves as an appropriate starting benchmark for various deep learning-based speech-enabled applications. For our experiments, we divide the dataset into 80% as training set and 20% for validation. The increased number of training examples for voice recognition results in better training performance of the model [29]. We performed random time shifting up to 20% as data-augmentation technique on training set. Audio files in Google speech commands dataset are 1 s long being sampled at 16 kHz. A time window of 25 ms is used for one frame to get 13 Mel-Frequency Cepstral Coefficients (MFCCs). MFCCs are well-tested audio features and extract most of the phoneme information from speech. Speech signal is essentially the sequence of different phonemes. The difference in amplitude between consecutive phonemes varies significantly. To capture this rich information of transitions, we use the difference between consecutive MFCCs of the whole signal which are known as delta features. Delta features are equivalent to derivative of the signal. We take difference once again of the consecutive delta features, and new features are then known as delta-delta (or double-delta) features. Delta and delta-delta features only capture the difference in consecutive phonemes; therefore it makes the feature map robust to the effect of white noise as well. For different speech commands the energy distribution of spectrum is different, so we use difference in energy value

Table 5.2 CASENet2 performance with different input features

Features type	Features dimension	Accuracy (%)
MFCC	13 × 40	94.11
MFCC, Delta	(13 + 13) × 40	94.89
MFCC, Delta, Double-Delta, Delta-Energy	(13 + 13 + 13 + 1) × 40	95.11

Table 5.3 CASENet hyperparameters count

Layer	Parameters	CASENet1	CASENet2	CASENet1-L
Conv. 1	Channels	64	32	32
	Kernel	5 × 5	5 × 5	5 × 5
Conv. 2	Channels	128	16	64
	Kernel	3 × 3	5 × 5	3×
Conv. 3	Channels	64	–	32
	Kernel	3 × 3	–	3 × 3
Dense	Nodes	128	100	100
Total no. of parameters		224,576	93,158	67,702

of spectrum for consecutive frames as our fortieth feature. The final shape of the feature map has a dimension of 40×40, where x-axis consists of time bins and y-axis corresponds to features related to frequencies. Utilizing the information from delta, double-delta, and delta-energy features improves the accuracy of the same model as listed in Table 5.2. The details of CASENet network architecture are explained in the following section.

2.4 CASENet and Its Comparison with State of the Art

CNN has delivered remarkable performance for image recognition tasks. They perform equally well for speech recognition given the speech signal is well expressed in form of features map. Various ML approaches have been proposed to improve the accuracy by increasing the network size, e.g., by increasing the number of parameters or going deeper in the network. However, quality and expressiveness of training data is equally important for high performance and accuracy. In our research, we have focused to extract maximum information from audio signals, which led to achieve high accuracy with significantly smaller model.

CASE is the abbreviation of "CNN Acceleration for Speech-Classification in Edge-Computing," and the CNN network which we have used is nicknamed as CASENet. We performed experiments with different hyperparameters of CNN and selected three different models to demonstrate our key observations and achievements. The hyperparameters for each CASENet are listed in Table 5.3. The maximum size of CASENet is chosen based upon resource capacity of targeted device. CASENet1 is our largest model but has 7× smaller number of parameters than CNN [27] and still achieves slightly higher accuracy. Similarly, CASENet1 achieves higher accuracy than RES15 [26] with 40× smaller number of operations. We reduced the number of parameters of CASENet1 up to 3× while maintaining above 95% accuracy and came up with a lighter version as CASENet1-L. Depthwise Separable CNN, DS-CNN [26], delivers good performance, but CASENet1-L achieves the same accuracy with 9× reduction in FLOPs and more than 2× reduction in the number of parameters. The number of operations of CASENet1-L is 9× smaller than EdgeSpeechNetD [24] but still achieves comparable accuracy. We also performed experiments by reducing the number of convolutional layers. CASENet2 has two convolutional layers and achieves lower accuracy than CASENet1-L even with a higher number of parameters. It demonstrates that the CNN layer improves the model performance with a smaller number of parameters but a larger number of FLOPS.

2.5 Hardware Introduction

The device used in this research is a custom-built Multiaccess Edge Computing (MEC) device named as M-KUBOS [30] which acts as edge server in our case. It has Xilinx Zynq Ultrascale+ XCZU19 System on Chip (SoC) processor on board. SoC on M-KUBOS has FPGA as programmable hardware and ARM processor as processing unit fabricated on the same chip. Moreover, SoC has on-chip memory for storing weights of the neural network logic which cuts short the reading and writing time of data into a separate memory while computations take place. SoC delivers high-end performance by leveraging the synergistic usage of processing unit and programmable hardware which communicate to each other through Advanced eXtensible Interface (AXI) as illustrated in Fig. 5.2. In addition to the on-chip memory, M-KUBOS has additional Ultra-RAM on board for storing large data which can be accessed by ARM processor and FPGA through Direct Memory Access (DMA). M-KUBOS has Linux operating system with Python production for Zynq devices (PYNQ) functionality. PYNQ framework in Linux enables easy development of applications based on Zynq devices by flexibly using developed hardware components in Python program. M-KUBOS acts as shh server and can be seamlessly accessed remotely.

High-Level Synthesis (HLS) tool by Vivado is used to design the CNN Logic in C language. Neural network logic can be synthesized and exported as Intellectual Property (IP) core to be flexibly used in hardware integration on Vivado Design Suite. Vivado Design Suite provides an immense amount of customization for SoC-based boards like M-KUBOS, where available hardware resources can be flexibly put together using Function Block Diagram (FBD)-based programming [31]. Finally integrated hardware design is synthesized, and bitstream file is burned into FPGA through Python-based application program in Linux operating system. Weights from the trained ML model are also transferred to on-chip memory, and the processing system uses FPGA as an accelerator for inference as shown in Fig. 5.2.

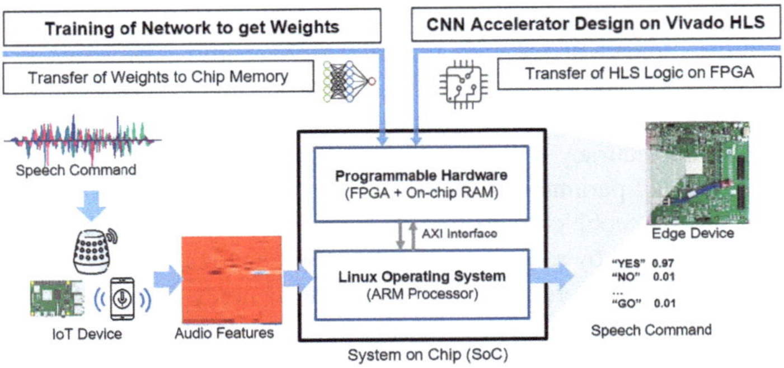

Fig. 5.2 Application framework for sound classification on edge device

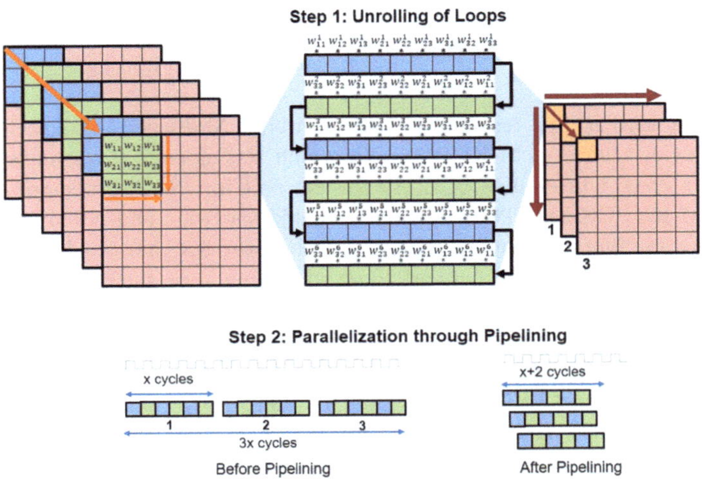

Fig. 5.3 Parallelization by loop unrolling followed by pipelining

2.6 *Accelerator Implementation*

A convolutional layer in neural network has six loops to compute activations for the next layer from input channels. As described in Algorithm 1, two innermost loops compute output for one kernel operation. The loop on top of kernel operation iterates over each channel of input layer. Looping over these kernel operations for every input channel followed by a summation returns a value corresponding to one output channel as shown in Fig. 5.3. These kernel operations in convolutional network provide an enormous amount of opportunity for parallelization. As each channel in output layer has different weights for all input channels, computations corresponding to output values for each output channel can be performed concurrently.

Achieving parallelization in convolutional operation is done in two steps, i.e., unrolling of for loops followed by pipelining. For example, in Fig. 5.3, three innermost loops are unrolled into a long vector which is responsible for outputting one value in an output channel. If the output channels are three, conventional approach would compute output activation corresponding to each channel in series manner. The simplest operation in CNN takes three clock cycles, i.e., reading, multiplying or adding, and writing. If computing a single value in one output channel takes x number of clock cycles, then three output values corresponding to each output channel would take 3× cycles. However, these independent operations can be scheduled to be executed in parallel with just one cycle delay for scheduling them as illustrated in Fig. 5.3.

Algorithm 1 Convolutional function

```
Require: input, weight, bias
Ensure: activation output
Convolution (input, weight, bias, output)
for w = 1 to output_channel_width do
    for h = 1 to output_channel_height do
        #pragma HLS PIPELINE
        for n = 1 to output_channels do
            #pragma HLS UNROLL
            for i = 1 to input_channels do
                for j = 1 to kernel_width do
                    for k = 1 to kernel_height do
                        output_whn += w_ijk × input_ijk
                        output_whn += bias_ijk
                    end for
                end for
            end for
                                          ▷ Above three loops unrolled into a vector.
        end for
                                          ▷ Above loop is pipelined for parallel processing.
    end for
end for
end Convolution
```

Vivado HLS tool allows to flexibly select the loops to be unrolled and parallelized. Algorithm 1 shows that we have unrolled three innermost loops of convolution function. The loop on top of three unrolled loops is pipelined and will execute its operation in parallel. Similarly, a more number of loops can be unrolled, and pipelining operation can be used for even higher loops. An important thing not to be here is that hardware resource required for parallel execution of unrolled loops would be significantly higher than doing one operation at a time. Theoretically all loops in convolution layer can be parallelized, given that there is no limitation on available resources. However, finding the optimal balance between the amount of parallelization and meeting the resource constraints is an important task [31]. Given the availability of computational resources in M-KUBOS, three to four loops are unrolled followed by pipelining in different convolutional layers of CASENet.

2.7 Experimental Setup and Results

Firstly, optimized CNN network, i.e., CASENet, is trained on Google Speech-Commands training dataset in PyTorch framework, and weights are obtained to transfer to chip memory before execution on M-KUBOS. Then HLS logic is synthesized on Vivado HLS, and a hardware bitstream file of neural network logic is transferred to M-KUBOS device for programming the FPGA hardware. A Python-based application program on operating system part of the SoC processor is made by utilizing PYNQ functionality. After programming FPGA hardware and transferring network weights to on-chip memory, Python-based application program sequentially takes the inputs and computes them by utilizing the on-chip FPGA accelerator. End-to-end latency is measured by executing base implementation of CASENet logic followed by its optimized version for CASENet1-L and CASENet2. Improvement in latency after using optimized version is illustrated in Fig. 5.4. We used Python-based time library to measure the end-to-end execution latency for speech classification inference.

In addition to our approach, some other works are being proposed for CNN acceleration on image classification tasks, where CNN implementation remains the same, but an audio features input matrix is replaced by image matrix. Fune [32], a CNN accelerator tool, achieves up to 2.14× speedup of CNN by providing optimal parameters search for hardware synthesis. ALAMO [33] is another approach which also explores CNN acceleration for the image recognition task by using an approach × speedup factor. Another proposal of CNN acceleration, Angle-Eye [34], achieved 6× as their best speedup factor by exploiting the potential of parallelization

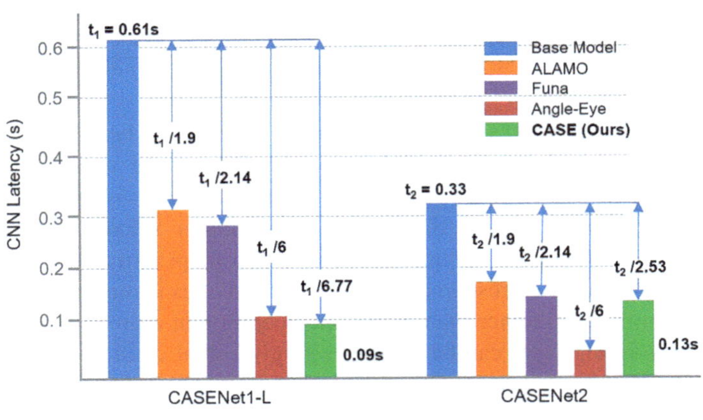

Fig. 5.4 Improvement in latency after parallelization in comparison to other methods

in CNN. Our approach in case of CASENet1-L has achieved an acceleration factor of 6.7×, which is higher than the previously mentioned approaches. An important observation in our experiment is that CASENet2 has lower latency without parallelization but still achieves lower acceleration factor than CASENet1. The reason is that three convolutional layers in CASENet1 provide more opportunity of parallelization than two convolutions in CASENet2. As convolution reduces the size of input, the lesser number of convolutional layers in network means more parameters before fully connected layer which has relatively lower potential of parallelization.

2.8 Summary

In this chapter, we demonstrated the use of SoC enabled edge server device as an accelerator for CNN-based speech classification. CASENet, a lightweight CNN network in terms of network parameters and a number of operations was proposed which achieved SOTA accuracy at significantly lower computational overhead. One of the key factors behind the success of CASENet was adopting a data-centric approach, which is to improve network performance by efficient feature extraction from input rather than increasing the network size. CASENet is the lightest CNN to achieve high performance which is comparable to significantly larger models. In the later part of the work, the design approach for CNN accelerator to be deployed on FPGA part of SoC processor was introduced. Our CNN accelerator achieved the maximum acceleration factor of 6.7× when compared with base implementation. Our future work includes:

- Extending the speech command classification to automatic speech recognition task
- Integration with IoT devices to make an end-to-end sound source localization, separation, and speech recognition framework
- Analyzing and improving the power consumption on edge device for making it energy efficient for practical applications

3 Sound Source Localization and Sound Source Separation

3.1 Background and Motivation

For SSL and SSS, the high computational costs and time consumption caused by a large number of matrix computations bring challenges for some practical applications [35]. Due to their universal compatibility and easy integration, most implementations realized SSL and SSS based on central processing units (CPUs) [9, 12, 36–42], which are typically suitable for small-scale data processing in embedded applications. For example, in drone audition [13], the lower CPU processing efficiency hinders real-time processing of large-scale microphone arrays on drones, limiting their ability to achieve higher quality SSL and SSS by increasing array size in extremely noisy environments. Additionally, inefficient processing leads to higher energy consumption and carbon footprint, which is unacceptable for edge computing that demands high performance and energy efficiency.

Although it is possible to deploy robot audition computations on cloud servers to handle data-intensive processing, the cloud-based setups have various drawbacks. Firstly, transmitting data between a large number of user devices and servers burdens the communication network and results in poor latency [43]. Secondly, sending personal data to the cloud for computation introduces data leakage and privacy risks [43, 44]. These challenges make it difficult for cloud-based methods to scale to an increasing number of devices. Therefore, implementing efficient computations for robot audition systems on local devices could be a better option, which brings new challenges. Many local devices are edge devices or embedded devices with limited computational resources [45, 46], necessitating the efficient deployment of robot audition systems by leveraging the resources of these devices themselves. In addition to CPUs, many embedded devices are equipped with graphics processing units (GPUs), and many edge computing devices are equipped with field-programmable gate arrays (FPGAs) to meet high-performance computing demands.

GPUs are processors specifically designed to accelerate graphics rendering and parallel computing tasks. Due to its highly parallel architecture, GPUs play an essential role in computer graphics and scientific computing, machine learning (ML), deep learning (DL), and large-scale data processing, which require high levels of parallelism. GPUs can process large-scale parallel computations at high speeds and have substantial data bandwidth.

FPGAs are programmable integrated circuits that allow users to configure and reprogram their logic structure at the hardware level. FPGAs have high parallel processing capabilities and low latency, making them suitable for customized computational tasks and applications with strict power consumption requirements. Generally, compared to the same generation of GPUs, FPGAs have bandwidth capabilities that are several orders of magnitude lower [47].

Based on their characteristics, GPUs are more suitable for applications that require intensive high-dimensional computations. In contrast, FPGAs are better suited for providing services in edge computing for simple auditory scenarios, such as indoor service robots, where small microphone arrays are sufficient, which have high demands for low latency and low power consumption.

For SSL and SSS, we target the robot audition platform HARK (Honda Research Institute Japan Audition for Robots with Kyoto University) [12], utilizing both GPUs and FPGAs for its implementations to meet the needs of different robot audition applications. Our contributions are:

1. We categorize the applications of robot audition systems and analyze how they can effectively leverage the inherent strengths of FPGAs and GPUs. By

identifying the optimal contexts for each type of hardware, we provide guidance for deploying implementations in environments that best utilize their respective capabilities.

2. We developed advanced parallel computing strategies for the Multiple Signal Classification based on Standard Eigenvalue Decomposition (SEVD-MUSIC)-based [48] SSL and the Geometric High-order Dicorrelation-based Source Separation (GHDSS)-based SSS [49] within HARK. These algorithms were implemented on NVIDIA GPUs using CUDA®, significantly reducing processing times and enabling real-time operation with a 60-channel microphone array. Evaluations demonstrate substantial speed enhancements, achieving a 23.3× speedup for SSL and 3.0× for SSS on a high-performance server configured with an NVIDIA A100 80 GB GPU, with 14.8× and 1.6× speedups, respectively, on the Jetson AGX Orin, representing embedded environments. These results underscore the effectiveness of the proposed GPU-based implementation in managing large-scale data flows in real time.
3. For edge computing applications, we adapted the SEVD-MUSIC-based SSL and GHDSS-based SSS modules of HARK for deployment on M-KUBOS [30] (PALTEK Corporation, Yokohama, Japan), an MEC FPGA MPSoC device. We significantly reduced transmission overhead by optimizing data transmission dynamics between the CPU and FPGA. Additionally, we strategically allocated the computational load to leverage the FPGA's parallel computing capabilities, achieving a 1.2× speedup for SSL and a 1.1× speedup for SSS with an eight-channel microphone array, along with a 1.1× improvement in overall energy efficiency. These results demonstrate the proposed FPGA-based implementation's ability to deliver high performance and energy efficiency in edge computing environments.

3.2 Related Work

Several studies have focused on efficiently implementing HARK on designated devices. Nakadai et al. [12] developed HARK-Embedded, an optimized version of HARK for ARM® processors, targeting embedded systems like smartphones and tablets. For four-channel audio data, this implementation achieved a performance level comparable to that of an Intel® Core™ i7 2640M processor on ARM® Cortex™ A9. However, according to our bottleneck analysis and experimental results, the CPU-based implementation fails to achieve real-time performance when processing large-scale audio data, even on a high-performance server CPU.

Hou et al. [50] implemented SSL module of HARK on an MEC FPGA MPSoC device M-KUBOS and significantly accelerated two matrix operation functions in the SSL module, which are named *AddCorrelation* and *NormalizeCorrelation*. Other parts of the SSL process were offloaded to an FPGA using automatic optimization provided by the high-level synthesis (HLS) tool. However, our bottleneck analysis reveals that these functions account for less than 2% of the total processing time of the SSL module. Furthermore, no significant acceleration was achieved for the SSL module as a whole. Despite deploying the module on an FPGA, CPU processing was found to be 3.5× faster than FPGA processing.

Qin et al. [51] deployed the SSS module of HARK on M-KUBOS for high-performance, low-power edge computing applications. Acceleration was achieved by optimizing the calculation of cost functions on an FPGA and utilizing nearly the maximum available resources to deploy SSS. However, in practical scenarios, SSL and SSS must operate concurrently, as SSL provides crucial output that guides SSS processing. To enable both modules to be efficiently accelerated on a single device, further optimization of SSS is necessary to maximize resource utilization and improve overall efficiency.

Lin et al. [52] developed an FPGA-based low-power acceleration for the SSL module by optimizing its most computationally intensive functions, designing parallel computing strategies, and deploying them on M-KUBOS. The study achieved low-power acceleration by setting the processing period to 1, the minimum value supported by HARK. However, this implementation did not account for data transmission overhead between the CPU and FPGA, requiring each frame of the Fourier-transformed sound signal to be transferred from the CPU to the FPGA. This overhead becomes a bottleneck with extended processing periods, such as 50, which are commonly used in practical applications due to higher transmission costs per processing period. Consequently, the implementation struggles to achieve effective acceleration under realistic conditions. Further optimization is necessary to reduce transmission overhead and enhance its performance in real-world scenarios.

Moreover, both [50] and [52] were limited to an eight-channel microphone array, constrained by the computing resources of the FPGA. Addressing the processing demands of large-scale microphone arrays requires leveraging the parallel computing capabilities of GPUs.

4 HARK

4.1 Processing Flow

Figure 5.5 illustrates an example of HARK's online batch processing flow with its Python interface, PyHARK, which is the basic architecture for HARK to perform SSL and SSS. It comprises four layers: the user program, pybind, HARK modules, and the operating system. The user program layer provides a programmable user interface and controls the processing flow. Pybind enables the Python user program to call HARK modules written in C++ and also carries the control flow of HARK modules. In the module layer, four modules are used for SSL and SSS: (1) *AudioStreamFromWave* reads multichannel audio data collected via a microphone array, (2) *MultiFFT* performs short-time Fourier transform (STFT) on the audio data, (3) *LocalizeMUSIC* performs MUSIC-based SSL, and (4) *GHDSS* utilizes

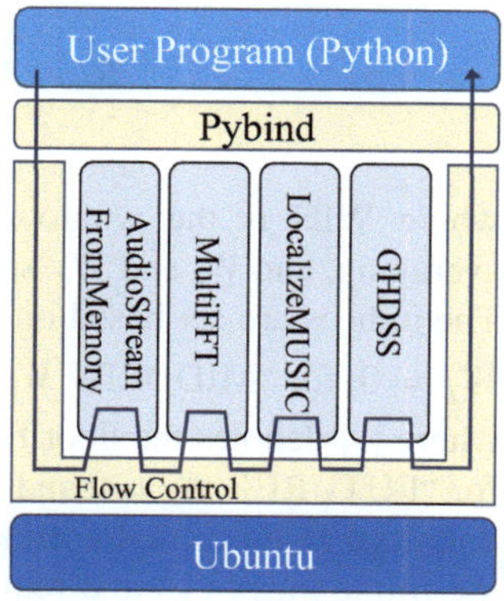

Fig. 5.5 Online batch processing flow of HARK for sound source localization (SSL) and sound source separation (SSS) tasks using its Python interface, PyHARK

the GHDSS algorithm to perform SSS. The operating system layer provides a runtime environment for the other three layers and handles system-level tasks such as process scheduling and hardware interaction.

4.2 *Sound Source Localization*

The SSL algorithm in HARK usually employs the SEVD-MUSIC method. This technique is widely used in various applications of HARK. SEVD-MUSIC operates as follows: The input audio signal, captured by an M-channel microphone array through the *AudioStreamFromWave*, is converted into the frequency domain by using the *MultiFFT* module to perform short-time Fourier transform (STFT), represented as $\mathbf{X}(\omega, f) \in \mathbb{C}^M$. Here, ω denotes the frequency bin and f signifies the frame index, defined as follows:

$$\mathbf{X}(\omega, f) = [X_1(\omega, f), X_2(\omega, f), \ldots, X_M(\omega, f)]^T . \tag{5.1}$$

LocalizeMUSIC takes $\mathbf{X}(\omega, f)$ as the input.

AddCorrelation computes an instantaneous correlation matrix ($\mathbf{R}_{\text{ins}}(\omega, f) \in \mathbb{C}^{M \times M}$) for each frequency bin as follows:

$$\mathbf{R}_{\text{ins}}(\omega, f) = \mathbf{X}(\omega, f)\mathbf{X}^H(\omega, f), \tag{5.2}$$

where $(\cdot)^H$ denotes the complex conjugate transpose operator. This matrix represents the correlation of signals across different channels of the microphone array at a single frame.

Subsequently, the *NormalizeCorrelation* function processes ($\mathbf{R}_{\text{ins}}(\omega, f)$ from *AddCorrelation*) to calculate the normalized correlation matrix $\mathbf{R}(\omega, f)$, defined as follows:

$$\mathbf{R}(\omega, f) = \frac{1}{\text{WIN}} \sum_{i=W_i}^{W_f} \mathbf{R}_{\text{ins}}(\omega, f+i), \tag{5.3}$$

where WIN is the window size, representing the number of frames used for averaging, and W_i and W_f specify the window's start and end indices, respectively. The indices are set based on the window type: For "PAST," $W_i = -\text{WIN} + 1$ and $W_f = 0$; for "MIDDLE," $W_i = -\left\lfloor \frac{\text{WIN}}{2} \right\rfloor$ and $W_f = \left\lfloor \frac{\text{WIN}}{2} \right\rfloor + (\text{WIN mod } 2) - 1$, where $\lfloor \cdot \rfloor$ denotes the floor function, which rounds down to the nearest integer; and for "FUTURE," $W_i = 0$ and $W_f = \text{WIN} - 1$. In an ideal scenario, $\mathbf{R}(\omega, f)$ should converge to the actual correlation matrix ($\mathbf{R}(\omega)$), which would be independent of the frame index (f) and require an infinite WIN. However, WIN is finite, and the estimated matrix is an approximation, rendering $\mathbf{R}(\omega, f)$ a positive semi-definite matrix that approximates the channel correlations within the microphone array.

Then, the *EVD* function performs SEVD on the normalized correlation matrix ($\mathbf{R}(\omega, f)$). This process involves decomposition ($\mathbf{R}(\omega, f)$) to distinguish between the signal and noise subspace. The decomposition can be expressed as follows:

$$\begin{aligned} \mathbf{R}(\omega, f) &= \mathbf{E}(\omega, f)\mathbf{\Lambda}(\omega, f)\mathbf{E}^H(\omega, f), \\ \mathbf{E}(\omega, f) &= [\mathbf{e}_1(\omega, f), \mathbf{e}_2(\omega, f), \cdots, \mathbf{e}_M(\omega, f)]^T, \\ \mathbf{\Lambda}(\omega, f) &= \text{diag}\left(\lambda_1(\omega, f), \lambda_2(\omega, f), \ldots, \lambda_M(\omega, f)\right), \end{aligned} \tag{5.4}$$

where $\mathbf{E}(\omega, f)$ indicates a matrix composed of eigenvectors and each vector ($\mathbf{e}_i(\omega, f)$) represents the i-th eigenvector associated with the frequency bin (ω) and frame index (f). $\mathbf{\Lambda}(\omega, f)$ is a diagonal matrix whose diagonal elements ($\lambda_i(\omega, f)$) are the eigenvalues corresponding to the eigenvectors in $\mathbf{E}(\omega, f)$. Each eigenvalue ($\lambda_i(\omega, f)$) quantifies the energy or power contribution of its corresponding eigenvector ($\mathbf{e}_i(\omega, f)$), and they are arranged in descending order from largest to smallest. Since the eigenvalue ($\lambda_i(\omega, f)$) corresponding to the eigenvector ($\mathbf{e_i}(\omega, f)$) has a positive relationship to the power of the sound source, the subspace composed of eigenvectors with larger eigenvalues is considered the sound source subspace. For example, if there are N_s sound sources, $\mathbf{e}_s = \left[\mathbf{e}_1(\omega, f), \mathbf{e}_2(\omega, f), \cdots, \mathbf{e}_{N_s}(\omega, f)\right]$ is the sound source subspace, while $\mathbf{e}_n = \left[\mathbf{e}_{N_{s+1}}(\omega, f), \cdots, \mathbf{e}_M(\omega, f)\right]$ is the noise subspace. In HARK, the solution is obtained through the Jacobi eigenvalue algorithm.

After that, *CalcAveragePower* calculates the power spectrum for each target direction using the MUSIC algorithm as follows:

$$\begin{aligned} \mathbf{P}(\theta, \omega, f) &= \frac{\left|\mathbf{H}^H(\theta, \omega)\mathbf{H}(\theta, \omega)\right|}{\sum_{i=N_s+1}^{M} \left|\mathbf{H}^H(\theta, \omega)\mathbf{e}_n(\omega, f)\right|}, \\ \mathbf{H}(\theta, \omega) &= [\mathbf{h}_1(\theta, \omega), \cdots, \mathbf{h}_M(\theta, \omega)]^T, \end{aligned} \tag{5.5}$$

where $\mathbf{H}(\theta, \omega) \in \mathbb{C}^M$ denotes the steering vector of the microphone array at a frequency of ω, representing the transfer function between the microphone array and a sound source originating from the direction of θ. Each element ($\mathbf{h}_i(\theta, \omega)$) is the transfer function of the i-th microphone. The transfer function ($\mathbf{H}(\theta, \omega)$) serves as a predetermined input to the module, encapsulating the array's characteristic response to sound from various directions across different frequencies. This process of *CalcAveragePower* uses eigenvectors of the noise subspace (e_n), from which the power spectrum is calculated in each direction. When the estimated direction accurately corresponds to the actual sound source direction, since the target signal is orthogonal to the noise subspace, the calculated power spectrum has a local maximum in this direction.

$\mathbf{P}(\theta, \omega, f)$ is then integrated in the dimension of the frequency as follows:

$$\bar{\mathbf{P}}(\theta, f) = \sum_{\omega=\omega_{\min}}^{\omega_{\max}} \mathbf{W}_{\Lambda}(\omega, f)\mathbf{W}_{\omega}(\omega, f)\mathbf{P}(\theta, \omega, f), \tag{5.6}$$

where $\mathbf{W}_{\Lambda}(\omega, f)$ is a weighting factor corresponding to eigenvalues, while $\mathbf{W}_{\omega}(\omega, f)$ is a weighting factor corresponding to frequency.

HARK ultimately identifies the direction of the sound source by searching for the peak value within the directional range of θ_{min} to θ_{max} of $\bar{\mathbf{P}}(\theta, f)$. This process is commonly conducted using the local maximum search.

4.3 Sound Source Separation

The GHDSS algorithm is described as follows. Let N denote the number of sound sources identified through SSL and M represent the number of microphones in the array. The complex spectrum vector of the mixed sources at the i-th frequency bin ($\mathbf{S}(\omega_i) \in \mathbb{C}^N$) is defined. Neglecting noise, the observation signal ($\mathbf{X}(\omega_i) \in \mathbb{C}^M$) recorded by the microphone array and transformed using STFT can be expressed as follows:

$$\mathbf{X}(\omega_i) = \mathbf{H}(\omega_i)\mathbf{S}(\omega_i), \tag{5.7}$$

where $\mathbf{H}(\omega_i) \in \mathbb{C}^{M\times N}$ represents the transfer function between sound sources and the microphone array at the i-th frequency (ω_i).

The separation in GHDSS is modeled as a linear process that estimates the separated sound sources ($\mathbf{Y}(\omega_i) \in \mathbb{C}^N$) from the observations ($\mathbf{X}(\omega_i)$). This is expressed as follows:

$$\mathbf{Y}(\omega_i) = \mathbf{W}(\omega_i)\mathbf{X}(\omega_i), \tag{5.8}$$

where $\mathbf{W}(\omega_i) \in \mathbb{C}^{N\times M}$ is the separation matrix, which is estimated based on high-order decorrelation and geometric constraints.

The GHDSS algorithm also optimizes the separation matrix by leveraging the high-order correlation matrix to eliminate high-order dependencies between signal sources. The high-order correlation matrix between the separated signals is defined as follows:

$$\mathbf{C}^{\phi(y)y}(\omega_i) = \mathbb{E}[\phi(\mathbf{Y}(\omega_i))\mathbf{Y}^H(\omega_i)], \tag{5.9}$$

where $(\cdot)^H$, $\mathbb{E}[\cdot]$, and $\phi(\cdot)$ denote the complex conjugation transpose, expectation operator, and nonlinear function, respectively. The nonlinear function ($\phi(\cdot)$) is expressed as follows:

$$\phi(\mathbf{Y}(\omega_i)) = \tanh(\sigma|\mathbf{Y}(\omega_i)|)\exp(j\angle(\mathbf{Y}(\omega_i))). \tag{5.10}$$

High-order decorrelation is achieved when all non-diagonal elements of $\mathbf{C}^{\phi(y)y}_{i,j}(\omega_i)$ are zero, so we can have the cost function for high-order decorrelation, expressed as follows:

$$J_1(\mathbf{W}(\omega_i)) = \sum_{i\neq j} \mathbf{C}^{\phi(y)y}_{i,j}(\omega_i). \tag{5.11}$$

Geometric constraints ensure that the spatial layout and physical transmission paths of the signals are taken into account during the signal separation process. Considering Equations (5.7) and (5.8), we can derive the following expression:

$$\mathbf{Y}(\omega_i) = \mathbf{W}(\omega_i)\mathbf{H}(\omega_i)\mathbf{S}(\omega_i). \tag{5.12}$$

To accurately restore the source signal ($\mathbf{S}(\omega_i)$), the following is required:

$$\mathbf{W}(\omega_i)\mathbf{H}(\omega_i) = \mathbf{I}. \tag{5.13}$$

Therefore, we can obtain the following cost function for geometric constraints:

$$J_2(\mathbf{W}(\omega_i)) = \|\mathbf{W}(\omega_i)\mathbf{H}(\omega_i) - \mathbf{I}\|^2. \tag{5.14}$$

When the number or position of sound sources detected by the SSL module remains unchanged between two consecutive frames, HARK updates the separation matrix. Otherwise, it saves the previous separation matrix and reinitializes a new one. The update of the separation matrix is based on the gradient descent method utilizing the complex gradient, expressed as follows:

$$\mathbf{W}^{(\mathbf{f}+1)}(\omega_i) = \mathbf{W}^{(\mathbf{f})}(\omega_i) - \left[\mu_1 \frac{\partial J_1}{\partial \mathbf{W}^{(\mathbf{f})*}}(\mathbf{W}^{(\mathbf{f})}(\omega_i)) + \mu_2 \frac{\partial J_2}{\partial \mathbf{W}^{(\mathbf{f})*}}(\mathbf{W}^{(\mathbf{f})}(\omega_i))\right], \tag{5.15}$$

$$\frac{\partial J_1}{\partial \mathbf{W}^*}(\mathbf{W}) = \left(\phi(\mathbf{Y})\mathbf{Y}^H - \text{diag}[\phi(\mathbf{Y})\mathbf{Y}^H]\right)\tilde{\phi}(\mathbf{WX})\mathbf{X}^H, \tag{5.16}$$

$$\frac{\partial J_2}{\partial \mathbf{W}^*}(\mathbf{W}) = 2\,(\mathbf{WH} - \mathbf{I})\,\mathbf{H}^H, \tag{5.17}$$

where μ_1 and μ_2 indicate step sizes, $(\cdot)^*$ is the complex conjugate operator, f represents the current time frame, $f+1$ denotes the next time frame after the update, and $\tilde{\phi}$ is the partial derivative of ϕ, which is defined as follows:

$$\tilde{\phi}(\mathbf{Y}) = [\tilde{\phi}(Y_1), \tilde{\phi}(Y_2), \cdots, \tilde{\phi}(Y_N)]^T, \tag{5.18}$$

$$\tilde{\phi}(Y_k) = \phi(Y_k) + Y_k \frac{\partial \phi(Y_k)}{\partial Y_k}, \tag{5.19}$$

while μ_1 and μ_2 are usually set to auto-adjust, calculated as follows:

$$\mu_1 = \frac{J_1(\mathbf{W})}{2\left\|\frac{\partial J_1}{\partial \mathbf{W}}(\mathbf{W})\right\|^2}, \tag{5.20}$$

$$\mu_2 = \frac{J_2(\mathbf{W})}{2\left\|\frac{\partial J_2}{\partial \mathbf{W}}(\mathbf{W})\right\|^2}, \tag{5.21}$$

and the initialization of $\mathbf{W}(\omega_i)$ is performed as follows:

$$\mathbf{W}(\omega_i) = \frac{\mathbf{H}^H(\omega_i)}{M}, \tag{5.22}$$

where M is the number of channels of the microphone array.

4.4 Methods and Implementations

4.4.1 Analysis of Bottlenecks

To identify performance bottlenecks in SSL and SSS of HARK, we measured the processing times of various parts within them on an NVIDIA Carmel ARM®v8.2 64-bit CPU.

Table 5.4 presents the processing time analysis of the internal functions within the SSL and SSS modules for each second of audio data obtained from a 60-channel microphone array. The table shows that a significant portion of the processing time

Table 5.4 The average processing time analysis of the internal functions within the SSL and SSS modules for each second of audio data obtained by a 60-channel microphone array, performed on an NVIDIA Carmel ARM® v8.2 64-bit CPU. This analysis sets the period to 50, meaning SSL processes every 50 frames (equivalent to 0.5 seconds of audio). The number of frequency bins is 73, and the system assumes the presence of two sound sources. *dUp* refers to the matrix operation function *dUpdateSeparationMatGHDSS*

Module	Functions		Time (ms)	Percentage	
SSL	Matrix operations	*AddCorrealtion*	508.49	1.81%	98.99%
		NormalizeCorrelation	4.66	0.02%	
		EVD	3160.53	11.27%	
		CalcAveragePower	24,096.88	85.89%	
	Non-matrix operation		282.45	1.01%	
Total (SSL)	–		28,053.01	100.00%	
SSS	Matrix operations	*dUp*	390.51	54.08%	
	Non-matrix operations		331.59	45.92%	
Total (SSS)	–		722.10	100.00%	

is spent on matrix operation functions, accounting for 98.99% in SSL and 54.08% in SSS. This indicates that matrix operations are the primary bottleneck in HARK.

4.4.2 Implementation of GPU Processing

To accelerate the processing of HARK and enable real-time processing with large-scale microphone arrays, as well as reducing preprocessing time for ML/DL, we leverage NVIDIA GPUs and CUDA® to optimize the bottleneck functions. To support microphone arrays of varying sizes and different frequency bin ranges, we utilize thread blocks of consistent size across matrices with different dimensions, dynamically adjusting the grid size based on the matrix dimensions.

The *AddCorrelation* and *NormalizeCorrelation* functions involve matrix operations with dimensions $\omega \times M \times M$, and every single element in the matrices can be calculated without depending on calculation results of other elements, so the degree of parallelism (DOP) is set to $\omega \times M \times M$.

EVD implements the Jacobi eigenvalue algorithm. For the i-th frequency bin (ω_i) of the current input frame, *EVD* iterates through all upper triangular elements ($R_{pq}(\omega_i, f)$) of the matrix ($R(\omega_i, f) \in \mathbb{C}^{M\times M}$) and applies rotational transformations to eliminate these elements, along with their symmetric counterparts ($R_{qp}(\omega_i, f)$). In this manner, the algorithm progressively approximates the eigenvalues of the matrix while simultaneously updating the eigenvector matrix. Each operation updates the rows and columns of the eigenvalue and eigenvector matrices corresponding to indices p and q. The updates of the elements in these rows or columns have no data dependencies on other elements in the same row or column, so M elements in a row or column can be updated in parallel. Moreover, eigenvalue decompositions for different frequency bins (ω) are mutually independent, allowing for parallel processing along the frequency dimension. Specifically, this implementation assigns a separate GPU block to each frequency bin (ω). Consequently, this implementation achieves a parallelism level of $\omega \times M$.

Furthermore, eigenvalue decompositions for different frequencies (ω) are mutually independent, allowing for parallel processing along the frequency dimension. Specifically, this implementation assigns a separate GPU block to each frequency (ω). Within each block, multiple threads (e.g., M threads) are employed to parallelize the updates across rows or columns of the $M \times M$ matrix. This design allows each thread to handle a different row or column, accelerating the application of rotational transformations. Consequently, this implementation achieves a parallelism level of $\omega \times M$.

The *CalcAveragePower* function, as described in Equations (5.5) and (5.6), involves operations such as multiplication, conjugate transpose, absolute value, and reduction (accumulation). All calculations, except for the final reduction step used to accumulate the power of each frequency bin across target directions—can be performed independently for each matrix element. Therefore, the maximum DOP is $\theta \times \omega \times M \times M$. However, for a 60-channel microphone array, the number of target directions (θ) typically exceeds 2500, and the frequency bin (ω) is usually set to 73 by default in the practical use of HARK. Therefore, if the maximum parallelism is implemented, the required threads exceed 675 million, significantly exceeding the maximum number of threads that even high-performance GPUs can execute simultaneously (up to 221,184 active threads for NVIDIA A100 GPU). Allocating an excessive number of threads can lead to increased scheduling overhead and diminished resource utilization efficiency. To optimize performance, we define the DOP as $\theta \times M \times M$, while maintaining a fixed grid size of θ and a thread block size of (32, 4). For instance, when θ is set to 2522, the maximum number of threads reaches 322,816.

The matrix operation function (*dUpdateSeparationMatGHDSS*), which handles the update of the separation matrix in SSS, the DOP is currently limited to the frequency bin level (ω) due to the complex data dependencies in the calculation.

Table 5.5 presents the allocation of thread blocks and grids for the aforementioned functions. For each function, the grid size is configured to cover the Degree of Parallelism (DOP), with the exception of *CalcAveragePower*. In CUDA

Table 5.5 Allocation of thread blocks and grids for different functions. M represents the number of channels, θ denotes the number of target directions in the array's transfer function, and ω specifies the number of frequency bins considered during processing

Function	Thread block size (block.x, block.y)	Grid size
AddCorrelation	(32, 4)	$(\lceil \frac{M}{block.x} \rceil, \lceil \frac{M}{block.y} \rceil, \omega)$
NormalizeCorrelation	(32, 4)	$(\lceil \frac{M}{block.x} \rceil, \lceil \frac{M}{block.y} \rceil, \omega)$
EVD	(64, 1)	$(\lceil \frac{M}{block.x} \rceil), \omega)$
CalcAveragePower	(32, 4)	$(\theta, 1, 1)$
dUpdateSeparationMatGHDSS	(32, 1)	$(\lceil \frac{\omega}{block.x} \rceil, 1)$

architecture, a warp is the fundamental unit of thread scheduling, comprising 32 threads that execute instructions concurrently. The thread block sizes are set to align with this warp structure. By configuring thread blocks with dimensions that are multiples or factors of the warp size, we ensure the optimal utilization of GPU resources and enhance execution efficiency.

4.4.3 FPGA Implementation of Sound Source Localization

According to the analysis presented in Table 5.4, the two most time-consuming functions within the SSL module are *EVD* and *CalcAveragePower*, which together account for 97.16% of the total processing time. Furthermore, since *AddCorrelation* and *NormalizeCorrelation* are executed for each time frame, deploying these functions on the FPGA would require transmitting every frame of the audio signal to the FPGA individually, introducing significant transmission overhead. Consequently, our approach focuses on implementing parallel computing strategies for the two most time-consuming functions in the SSL module while retaining the other functions on the CPU.

Figure 5.6 shows the architecture of the implementation of the SSL module. *EVD* and *CalcAveragePower* are parallelized with a DOP of 12 by deploying 12 corresponding units. This value of 12 represents the maximum achievable DOP, given the resource constraints of the FPGA. Other parts of the module are deployed on the CPU to avoid transmitting the data from every frame between the CPU and FPGA. Algorithm 2 depicts the detailed implementation methods of *EVD* and *CalcAveragePower*.

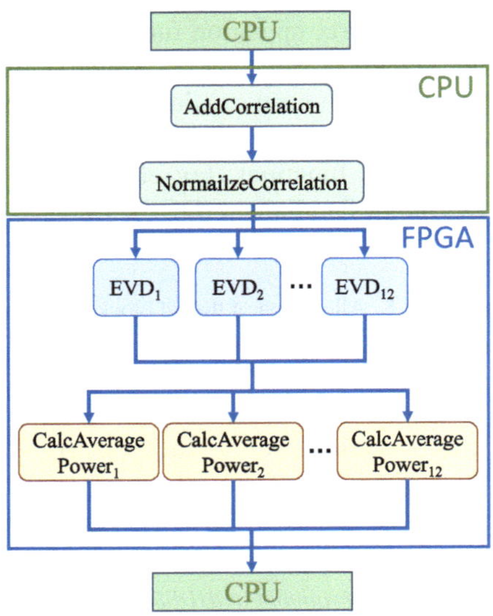

Fig. 5.6 Architecture of the SSL implementation

Algorithm 2 Parallel computing of *EVD* and *CalcAveragePower* in HLS

Require: Initialization:
1: $\omega_{min} \leftarrow$ the lowest frequency index
2: $\omega_{max} \leftarrow$ the highest frequency index
3: $M \leftarrow$ the number of microphones
4: $N_{\text{IP}} \leftarrow$ the number of frequency bins per IP core
5: $N_{bin} \leftarrow$ the total length of the frequency bins
6: $N_{pos} \leftarrow$ the number of target directions
7: $\text{EVD}_1, \ldots, \text{EVD}_{12}$ for parallelized *EVD* functions
8: $\text{CalcAveragePower}_1, \ldots, \text{CalcAveragePower}_{12}$ for parallelized *CalcAveragePower* functions
9: **for** $i = \omega_{min}$ **to** ω_{max} **step** 12 **do**
10: // Prepare data for parallelized *EVD* functions
11: $R_j \leftarrow [\quad]$ // Initialize as an empty array
12: **for** $j = 1$ **to** 12 **do**
13: Append $\mathbf{R}(i + j - 1, f)$ to R_j
14: **end for**
15: // N parallel computation of *EVD*
16: $[\Lambda_1, E_1] \leftarrow \text{EVD}_1(R_1)$
17: $[\Lambda_2, E_2] \leftarrow \text{EVD}_2(R_2)$
18: $\vdots$
19: $[\Lambda_{12}, E_{12}] \leftarrow \text{EVD}_{12}(R_{12})$
20: **for** $k = 0$ **to** $N_{pos} - 1$ **do**
21: // Prepare data for parallelized *CalcAveragePower*
22: $H_{k,j} \leftarrow [\quad]$ // Initialize as an empty array
23: **for** $j = 1$ **to** 12 **do**
24: Append $\mathbf{H}(k, i + j - 1)$ to $H_{k,j}$ // Burst transmission
25: **end for**
26: // N parallel computation of *CalcAveragePower*
27: $P_1 = \text{CalcAveragePower}_1(H_{k,1}, \Lambda_1, E_1)$;
28: $P_2 = \text{CalcAveragePower}_2(H_{k,2}, \Lambda_2, E_2)$;
29: $\vdots$
30: $P_{12} = \text{CalcAveragePower}_{12}(H_{k,12}, \Lambda_{12}, E_{12})$;
31: $P[k] = \sum_i^N P_i$
32: **end for**
33: **end for**

The "for" loop in line 9 is to traverse all frequency bins. Lines 10–14 prepare data for parallelized *EVD* operations. $\mathbf{R}(\omega, f)$ is divided into 12 blocks with different frequency bins as $[R_1, \cdots, R_{12}]$.

Lines 15–19 execute parallel *EVD* operations along the frequency bin dimension, aligning with our FPGA implementation strategy. The DOP is set to 12, representing the maximum achievable under the resource constraints. In this configuration, 12 correlation matrices $[R_1, \cdots, R_{12}]$ can be processed simultaneously, producing 12 sets of eigenvalues and eigenvectors ($[\Lambda_1, \cdots, \Lambda_{12}]$ and $[E_1, \cdots, E_{12}]$, respectively).

The "for" loop in line 20 is to traverse all target directions. Lines 21–25 prepare data for parallelized *CalcAveragePower* operations, that is, loading transfer functions for each target direction (N_{pos}) are prepared for the 12 frequency bins considered in former *EVD* calculation. Since the transfer functions are stored continuously in the frequency bin dimension, we utilize burst transmission instead of DDR4 memory access to reduce transmission overhead.

Lines 27–31 compute the MUSIC power spectra in parallel, utilizing the results from the preceding *EVD* operations. Parallelism is implemented along the target direction dimension, with a DOP of 12, following the same FPGA implementation strategy and resource constraints applied to *EVD*. The i-th MUSIC power spectrum (P_i) for the θ_k direction is calculated using Λ_i, E_i, and $H_{k,j}$.

In the final step, line 32 accumulates the power values from the N MUSIC power spectra for each direction (θ_k), generating the overall power spectrum ($P[k]$).

4.4.4 FPGA Implementation of Sound Source Separation

In the FPGA implementation of SSS, we target the matrix operation function (*dUpdateSeparationMatGHDSS* (*dUp*)), which updates the separation matrix and deploys it as an IP core.

Since the range of frequency bins is an adjustable parameter in HARK, we cannot utilize the loop *UNROLL* pragma in Vitis HLS to parallelize *dUp* across the frequency bin dimension as the proposed GPU implementation. Therefore, parallel computation is implemented by applying the loop *UNROLL* pragma in the appropriate "for" loops within *dUp*. In the steps involving the calculation of the separation matrix ($\mathbf{W}(\omega_i) \in \mathbb{C}^{N\times M}$), the computations require the expansion of $\mathbf{W}$ along both the N and M dimensions. Since N is adjustable and M is fixed at 8, we place the loop expanding along the N dimension as the outer loop, while the loop along the M dimension is the inner loop. We apply the loop *UNROLL* pragma to optimize the loops with fixed cycles. Similarly, we also apply the loop *UNROLL* pragma to parallelize other loops with a fixed iteration count.

4.5 Evaluation

4.5.1 Metrics and Measurement Methods for Computational Efficiency

To evaluate the performance and energy efficiency of our GPU-based implementation, we measured processing times and compared them with those of the CPU-based implementation. We conducted this analysis to determine whether our approach meets real-time processing requirements and assess its potential for enhancing efficiency when HARK is a preprocessing step in machine learning tasks.

For the FPGA-based implementation, we measured processing times and power consumption and compared these results with the CPU-based implementation to evaluate both performance and energy efficiency comprehensively.

We recorded processing times using timers embedded in the code of each module and function, logging the results to text files for further analysis with Python scripts. We conducted power measurements using a specialized setup with monitoring capabilities, including an LTM4677 analog device for DC/DC conversion. This setup connects to a Windows PC via a DC2086A Powered Programming Adapter and a USB-connected DC1613A I^2C controller. We recorded and exported data in CSV format using LTPowerPlay software. These measurements capture the average power consumption of M-KUBOS during SSL and SSS tasks in the HARK program, providing a detailed understanding of the system's energy efficiency.

4.5.2 Metrics and Measurement Methods for Calculation Accuracy

When migrating computations from CPU-based implementations to GPU or FPGA platforms, verifying calculation accuracy is crucial to ensure consistent and reliable results. Differences in hardware architecture, numerical precision, and parallel computation methods can introduce discrepancies in the output. We conducted the following evaluations to assess the accuracy of the proposed GPU- and FPGA-based implementations.

For SSL, we compared the sound source positions detected by the GPU- and FPGA-based implementations in each processing period with those of the CPU-based implementation on each device to ensure consistency.

For SSS, we calculated the Relative Root Mean Squared Error (RRMSE) of the separation matrices generated by the GPU and FPGA implementations relative to the CPU-based results on each device to quantify computational discrepancies. The RRMSE is calculated as follows:

$$\mathrm{RRMSE} = \sqrt{\frac{\sum_{k=1}^{K}\sum_{i=1}^{N}\sum_{j=1}^{M}\left|W_{\mathrm{test},ij}^{(k)} - W_{\mathrm{ref},ij}^{(k)}\right|^2}{\sum_{k=1}^{K}\sum_{i=1}^{N}\sum_{j=1}^{M}\left|W_{\mathrm{ref},ij}^{(k)}\right|^2}},$$

where k is the index of the separation matrices, W_{test} is the separation matrix computed by the GPU- or FPGA-based implementations, and W_{ref} is the reference matrix calculated by the CPU-based implementation.

4.5.3 Experimental Settings

For the GPU-based implementation evaluations, we conducted tests on two distinct types of devices: a high-performance server featuring an NVIDIA A100 80GB GPU and AMD EPYC™ 7352 CPUs, simulating applications in machine learning scenarios, and the Jetson AGX Orin 32GB, an embedded GPU device powered by ARM®Cortex™-A78AE v8.2 64-bit CPUs, representing applications in robots or embedded devices. This diverse hardware setup also enables us to assess the

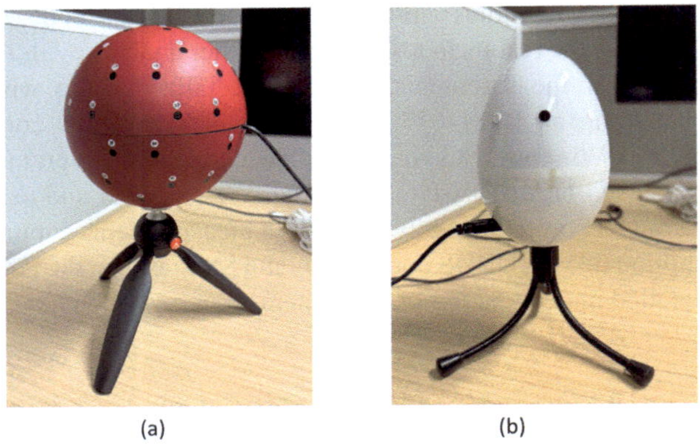

Fig. 5.7 Microphone arrays utilized in evaluations (**a**) C60. (**b**) TAMAGO

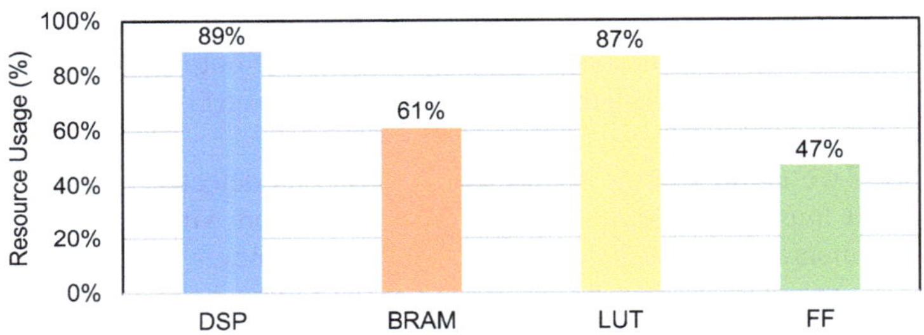

Fig. 5.8 Resource usage of FPGA-based implementation

effectiveness of the implementation across different scales of GPU capabilities. The evaluations were conducted using a 60-channel microphone array named C60, as shown in Fig. 5.7a, with both CPU and GPU processing performed in single-precision floating-point format.

In the configuration of HARK, we used a default setup where the sampling rate was set at 16 kHz, covering a frequency range of 500 Hz–2800 Hz across 73 frequency bins. The window type was configured as "PAST" with a size of 50, and the processing period was established at 50 time frames, equivalent to 0.5 s of audio. This setup is typical for practical applications. The audio analysis accounts for two sound sources, and 2522 discrete target directions cover a full 360-degree spherical range.

For the FPGA-based implementation evaluations, we conducted evaluations on M-KUBOS, while processing was performed in single-precision floating-point format. We deployed both an SSL module and an SSS module on M-KUBOS, which nearly reached the FPGA's capacity limits. The resource utilization of the FPGA is shown in Fig. 5.8.

For the setup of HARK, the sampling rate was 16 kHz, while the frequency bin was 96. The target microphone array, TAMAGO, exhibited in Fig. 5.7b, is an eight-

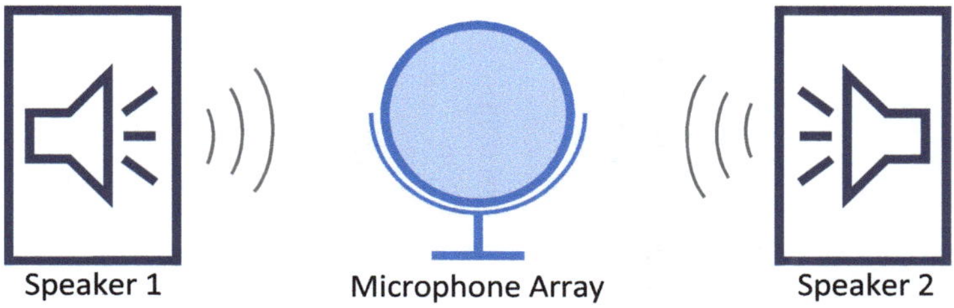

Fig. 5.9 Experimental setup for audio data acquisition

channel microphone array that provides full 360-degree coverage on a plane divided into 72 discrete target directions, which are fixed in the implementation. As for the GPU settings, the processing period and window size were configured to 50 time frames, and the window type was set to "PAST."

The audio data used in the experiments were collected in advance using the setup illustrated in Fig. 5.9. Two speakers were placed on either side of the microphone array, serving as two independent sound sources. These speakers played speech signals while the microphone array recorded the audio. The recordings were performed using C60 and TAMAGO microphone arrays and saved in a multichannel `.wav` format. Each recording lasted for 120 seconds.

4.5.4 Performances of GPU Implementation

Table 5.6 presents the evaluation results of the GPU-based implementation. The results show that CPU-based processing predominantly consumes time in matrix operations, hindering real-time processing capabilities for both the A100 server and the Jetson AGX Orin when using the 60-channel microphone array. In contrast, the GPU-based implementation markedly decreases the matrix operation times on both devices. Specifically, on the A100 server, the GPU achieves speedups of 41.5× and 16.0× for matrix operations in SSL and SSS, respectively. The total speedups for the SSL and SSS modules are 23.3× and 3.0×, respectively. On the Jetson AGX Orin, the speedups for matrix operations are 18.6× and 1.9× for SSL and SSS, respectively, with overall module speedups of 14.8× and 1.6×. Importantly, as shown in Fig. 5.10, GPUs facilitate real-time processing on both devices by ensuring that the total processing time remains well below 1 second for each second of audio data. This also provides ample capacity to achieve real-time processing of potential subsequent machine learning and deep learning (ML/DL) tasks. These results underscore the effectiveness of the proposed generalizable GPU-based implementation of HARK.

Table 5.6 The average processing times of SSL and SSS on central processing units (CPUs) and graphics processing units (GPUs) of different devices for each second of 60-channel audio, presented in milliseconds (ms). Results include the processing time of matrix operations (ops) and non-matrix operations in SSL and SSS

		SSL			SSS		
Device	Processor	Matrix ops	Non-matrix ops	Total (SSL)	Matrix ops	Non-matrix ops	Total (SSS)
A100 Server	CPU	2477.05	47.50	2524.55	201.46	75.90	277.36
	GPU	59.73	48.45	108.58	12.60	78.77	91.37
Jetson AGX Orin	CPU	3561.18	57.27	3618.45	184.37	94.10	278.47
	GPU	191.60	52.47	244.07	94.59	83.49	178.08

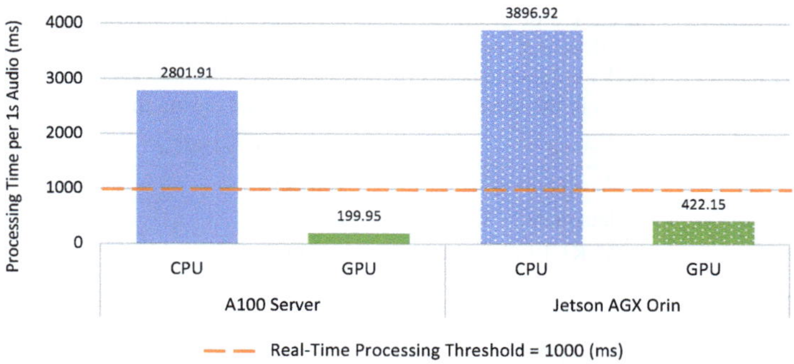

Fig. 5.10 Total average processing time for SSL and SSS for each second of 60-channel audio on different processors. Bars below the threshold indicate real-time capability, while bars above the threshold do not meet real-time requirements

Table 5.7 The average processing time, average power, and power consumption of SSL and SSS with different implementations on each second of the eight-channel audio

Implementation	Task	Time (ms)	Average power (W)	Power consumption (mJ)
CPU	SSL	77.33	5.94	459.34
	SSS	558.15		3315.41
	Total (SSL+SSS)	635.48		3774.75
FPGA	SSL	64.97	6.21	403.46
	SSS	492.42		3057.93
	Total (SSL+SSS)	557.39		3461.39
FPGA (all matrix operations deployed)	SSL	661.87	6.21	4110.21

4.5.5 Performances of FPGA Implementation

Table 5.7 presents the evaluation results for the FPGA-based implementation, focusing on processing time and power consumption. The results demonstrate significant reductions in SSL processing time and power consumption compared to FPGA implementations that perform all matrix operations. These improvements were achieved by optimizing the architecture to minimize transmission overhead. Compared to CPU-based implementations, the FPGA solution achieved speedups of 1.2× for the SSL module and 1.1× for the SSS module. The power consumption figures include both CPU and FPGA, as the SoC integrates these components on a single chip. Although the average power increased by 3.65%, the reduction in processing time led to a lower overall energy consumption for the FPGA implementation compared to CPU-based processing. Specifically, the FPGA implementation realized 1.1× higher energy efficiency for both SSL and SSS, highlighting the effectiveness of the proposed approach.

Table 5.8 Calculation accuracy of proposed implementations. Consistency represents the percentage of outputted sound source positions from proposed implementations that match the outputs of the CPU-based implementation. RRMSE indicates the average relative root mean squared error of all elements in the separation matrices outputted during the process

Implementation	Device	Consistency (SSL)	RRMSE (SSS)
GPU	A100 Server	100.00%	2.61×10^{-5}
	Jetson AGX Orin	100.00%	2.59×10^{-5}
FPGA	M-KUBOS	100.00%	9.26×10^{-6}

4.5.6 Calculation Accuracy

Table 5.8 presents the evaluation results of calculation accuracy for the proposed GPU- and FPGA-based implementations.

For SSL, the consistency metric represents the percentage of sound source positions detected by the GPU and FPGA implementations that match the CPU implementation outputs. As shown in Table 5.8, all tested devices achieved a consistency of 100.00%, confirming that the proposed implementations reliably reproduce CPU-based SSL results.

For SSS, the RRMSE values demonstrate high numerical accuracy across all platforms. Specifically, the GPU-based implementation achieved an RRMSE of 2.61×10^{-5} on the A100 server and 2.59×10^{-5} on the Jetson AGX Orin, while the FPGA-based implementation achieved an RRMSE of 9.26×10^{-6} on M-KUBOS. These results indicate that the proposed implementations produce separation matrices with negligible errors, comparable to CPU results in practical applications.

Overall, the evaluation confirms that the GPU and FPGA implementations maintain high calculation accuracy, ensuring their reliability for practical use.

5 Summary and Future Work

This work categorizes applications of robot audition systems, discusses the devices they are suitable for, and presents efficient implementations of these systems, with a specific focus on deploying HARK, an open-source software platform designed for robot audition. Our work focuses on implementations using GPUs and FPGAs, providing efficient solutions for data-intensive scenarios and edge computing environments. The generalizable GPU-based implementation substantially reduced processing times, enabling real-time operation of SSL and SSS with a 60-channel microphone array. This was evidenced by remarkable speedups achieved on GPU devices of varying capabilities: 23.3× for SSL and 3.0× for SSS on an NVIDIA A100 80 GB GPU server and 14.8× for SSL and 1.6× for SSS on a Jetson AGX Orin. Furthermore, the proposed FPGA-based implementation, designed using HLS, was tailored for edge computing environments. With an eight-channel microphone array, this approach resulted in a 1.2× speedup for SSL and a 1.1× speedup for SSS, along with 1.1× enhanced energy efficiency on the M-KUBOS, an MEC FPGA MPSoC device.

Through the proposed implementations, robot audition systems can rapidly process data from large-scale microphone arrays and achieve high-performance, low-power processing in edge computing environments, making them more adaptable and efficient for real-world applications. For example, the GPU-based implementation enables rescue robots or drones equipped with embedded systems like Jetson to utilize large-scale microphone arrays for precise, real-time SSL and SSS, thereby facilitating faster detection of survivors. Meanwhile, the FPGA-based implementation provides high-performance processing for private users while reducing electricity costs and carbon footprints in edge computing environments.

However, the extra power consumption introduced by GPU processing may affect the battery life of devices such as drones and robots. The current implementation has not yet been tested on these devices' onboard platforms, and the extent of the impact remains unclear. Therefore, future in-depth research and field testing are required to evaluate the specific impact of GPU processing on device battery life and to optimize system design to balance performance with energy consumption.

On the other hand, the proposed GPU-based implementation cannot achieve stream processing on a 60-channel microphone array. Stream processing must handle each audio frame in real time, ensuring the system can still provide reliable real-time audio processing capabilities in complex and dynamic application environments. In the current implementation, audio frames are processed at determined intervals, inevitably introducing a delay. Stream processing is an ideal solution in scenarios with frequently changing sound sources and the need for rapid responses. Future work will focus on improving the existing algorithms to enhance efficiency and further transition to a stream processing mode to increase the system's response speed and processing accuracy.

For the FPGA-based implementation, resource limitations make it challenging to increase DOP on the FPGA, thereby restricting further reductions in processing time and improvements in energy efficiency. Therefore, another future work should involve algorithm optimization tailored to FPGA resources to utilize these resources and achieve more efficient implementations.

6 Conclusion of This Section

Robot audition technology, encompassing Automatic Speech Recognition (ASR), Sound Source Localization (SSL), and Sound Source Separation (SSS), plays a pivotal role in equipping robots and smart devices with human-like auditory capabilities. Effective ASR implementations enable robust voice command interfaces, while SSL and SSS techniques facilitate real-time detection and isolation of sound sources in diverse, often noisy, environments. Accelerating these processes through GPUs and FPGAs, as well as implementing lightweight design, significantly reduces latency and energy consumption, making such solutions ideal for both large-scale and edge computing scenarios. Future advancements will focus on refining algorithms for better resource utilization, further reducing latency through real-time stream processing, and exploring specialized hardware accelerators, thereby expanding the possibilities for high-performance and energy-efficient robot audition systems.

References

1. J. Blauert, *Communication Acoustics* (Springer, New York, 2005), p. 1. https://doi.org/10.1007/b139075
2. R.N. Albustanji, S. Elmanaseer, A.A.A. Alkhatib, Robotics: five senses plus one—an overview. Robotics **12**, 68 (2023). https://doi.org/10.3390/robotics12030068
3. D. Moore, W. Ju, Sound as implicit influence on human-robot interactions, in *Proceedings of the HRI '18, Companion of the 2018 ACM/IEEE International Conference on Human-Robot Interaction* (New York, 2018), pp. 311–312. https://doi.org/10.1145/3173386.3176918
4. T.J. Tsai, A. Stolcke, M. Slaney, A study of multimodal addressee detection in human-human-computer interaction. IEEE Trans. Multimedia **17**, 1550–1561 (2015). https://doi.org/10.1109/TMM.2015.2454332
5. J.-C. Wang et al., Robust environmental sound recognition for home automation. IEEE Trans. Autom. Sci. Eng. **5**(1), 25–31 (2008). https://doi.org/10.1109/TASE.2007.911680
6. B.J. Zhang, N.T. Fitter, Nonverbal sound in human-robot interaction: a systematic review. J. Hum.-Robot Interact. **12**, 1–46 (2023). https://doi.org/10.1145/3583743
7. H. Pelikan et al., Sound in human-robot interaction, in *Proceedings of the HRI '21, Companion of the 2021 ACM/IEEE International Conference on Human-Robot Interaction* (New York, 2021), pp. 706–708. https://doi.org/10.1145/3434074.3444871
8. H.G. Okuno, K. Nakadai, Robot audition: its rise and perspectives, in *2015 IEEE International Conference on Acoustics, Speech and Signal Processing (ICASSP)* (2015), pp. 5610–5614. https://doi.org/10.1109/ICASSP.2015.7179045
9. H.W. Löllmann et al, Microphone array signal processing for robot audition, in *Proceedings of the 2017 Hands-free Speech Communications and Microphone Arrays (HSCMA)* (San Francisco, 2017), pp. 51–55. https://doi.org/10.1109/HSCMA.2017.7895560
10. K. Kumatani, J. McDonough, B. Raj, Microphone array processing for distant speech recognition: from close-talking microphones to far-field sensors. IEEE Signal Process. Mag. **29**, 127–140 (2012). https://doi.org/10.1109/MSP.2012.2205285
11. I. McCowan et al., *Towards Computer Understanding of Human Interactions*. Lecture Notes in Computer Science, vol. 3361 (2003), pp. 235–251. https://doi.org/10.1007/978-3-540-39863-9%5C_18
12. K. Nakadai, H.G. Okuno, T. Mizumoto, Development, deployment and applications of robot audition open source software HARK. J. Robot. Mechatron. **29**, 16–25 (2017). https://doi.org/10.20965/jrm.2017.p0016
13. K. Nakadai, H.G. Okuno, Robot audition and computational auditory scene analysis. Adv. Intell. Syst. **2**(9), 2000050 (2020). https://doi.org/10.1002/aisy.202000050
14. K. Ajay, N. Srinidhi, K. Ramaswamy, Fully responsive image and speech detection artificial yankee (FRIDAY): human assistant. SN Comput. Sci. **2**(4), 243 (2021)
15. J. Novoa et al., Automatic speech recognition for indoor HRI scenarios. J. Hum.-Robot Interact. **10**(2) (2021). https://doi.org/10.1145/3442629
16. M.C. Bingol, O. Aydogmus, Performing predefined tasks using the human–robot interaction on speech recognition for an industrial robot. Eng. Appl. Artif. Intell. **95**, 103903 (2020). ISSN: 0952-1976. https://doi.org/10.1016/j.engappai.2020.103903

17. C.M.J. Galangque, S.A. Guirnaldo, Speech recognition engine using ConvNet for the development of a voice command controller for fixed wing unmanned aerial vehicle (UAV), in *2019 12th International Conference on Information Communication Technology and System (ICTS)* (2019), pp. 93–97. https://doi.org/10.1109/ICTS.2019.8850961
18. M. Vukovic, M. Stolar, M. Lech, Cognitive load estimation from speech commands to simulated aircraft. IEEE/ACM Trans. Audio Speech Lang. Process. **29**, 1011–1022 (2021). https://doi.org/10.1109/TASLP.2021.3057492
19. C. Zonios, V. Tenentes, Energy efficient speech command recognition for private smart home IoT applications, in *2021 10th International Conference on Modern Circuits and Systems Technologies (MOCAST)* (2021), pp. 1–4. https://doi.org/10.1109/MOCAST52088.2021.9493392
20. J. Mayer, IoT architecture for home automation by speech control aimed to assist people with mobility restrictions, in *Proceedings on the International Conference on Internet Computing (ICOMP)* (2017), pp. 92–97
21. G. Dinelli et al., Advantages and limitations of fully on-chip CNN FPGA-based hardware accelerator, in *2020 IEEE International Symposium on Circuits and Systems (ISCAS)* (2020), pp. 1–5. https://doi.org/10.1109/ISCAS45731.2020.9180867
22. O. Rybakov et al., Streaming keyword spotting on mobile devices, in *INTERSPEECH* (2020). https://doi.org/10.21437/Interspeech.2020-1003
23. S. Yang et al., EdgeRNN: a compact speech recognition network with spatio-temporal features for edge computing. IEEE Access **8**, 81468–81478 (2020). https://doi.org/10.1109/ACCESS.2020.2990974
24. Z. Qiu Lin, A.G. Chung, A. Wong. *EdgeSpeechNets: Highly Efficient Deep Neural Networks for Speech Recognition on the Edge* (2018). arXiv: 1810.08559 [eess.AS]
25. R. Tang, J. Lin, Deep residual learning for small-footprint keyword spotting, in *2018 IEEE International Conference on Acoustics, Speech and Signal Processing (ICASSP)* (2018), pp. 5484–5488. https://doi.org/10.1109/ICASSP.2018.8462688
26. Y. Zhang et al., *Hello Edge: Keyword Spotting on Microcontrollers* (2018). arXiv: 1711.07128 [cs.SD]
27. A. Soliman, S. Mohamed, I.A. Abdelrahman, Isolated word speech recognition using convolutional neural network, in *2020 International Conference on Computer, Control, Electrical, and Electronics Engineering (ICCCEEE)* (2021), pp. 1–6. https://doi.org/10.1109/ICCCEEE49695.2021.9429684
28. P. Warden, *Speech Commands: A Dataset for Limited-Vocabulary Speech Recognition* (2018). arXiv: 1804.03209 [cs.CL]
29. A. Nur Çayır, T.S. Navruz, Effect of dataset size on deep learning in voice recognition, in *2021 3rd International Congress on Human-Computer Interaction, Optimization and Robotic Applications (HORA)* (2021), pp. 1–5. https://doi.org/10.1109/HORA52670.2021.9461395
30. T. Inage et al., M-KUBOS/PYNQ cluster for multi-access edge computing, in *Proceedings of the 2021 Ninth International Symposium on Computing and Networking (CANDAR)* (Matsue, 2021), pp. 95–101. https://doi.org/10.1109/CANDAR53791.2021.00020
31. L. Stornaiuolo, M. Santambrogio, D. Sciuto, On how to efficiently implement deep learning algorithms on PYNQ platform, in *2018 IEEE Computer Society Annual Symposium on VLSI (ISVLSI)* (2018), pp. 587–590. https://doi.org/10.1109/ISVLSI.2018.00112
32. Q. Xiao, Y. Liang, Fune: an FPGA tuning framework for CNN acceleration. IEEE Design Test **37**(1), 46–55 (2020). https://doi.org/10.1109/MDAT.2019.2908549
33. Y. Ma et al., ALAMO: FPGA acceleration of deep learning algorithms with a modularized RTL compiler. Integration **62**, 14–23 (2018). ISSN: 0167-9260. https://doi.org/10.1016/j.vlsi.2017.12.009
34. K. Guo et al., Angel-eye: a complete design flow for mapping CNN onto embedded FPGA. IEEE Trans. Comput.-Aided Design Integrated Circuits Syst. **37**(1), 35–47 (2018). https://doi.org/10.1109/TCAD.2017.2705069
35. R. Guo, Z. Luo, M. Li, A survey of optimization methods for independent vector analysis in audio source separation. Sensors **23**, 493 (2023). https://doi.org/10.3390/s23010493

36. F. Michaud et al., Spartacus attending the 2005 AAAI conference. Auton. Robots **22**, 369–383 (2007). https://doi.org/10.1007/s10514-006-9014-7
37. J.-M. Valin, F. Michaud, J. Rouat, Robust localization and tracking of simultaneous moving sound sources using beamforming and particle filtering. Robot. Auton. Syst. **55**, 216–228 (2007). https://doi.org/10.1016/j.robot.2006.08.004
38. S. Yamamoto et al., Real-time robot audition system that recognizes simultaneous speech in the real world, in *Proceedings of the 2006 IEEE/RSJ International Conference on Intelligent Robots and Systems* (Beijing, 2006), pp. 5333–5338. https://doi.org/10.1109/IROS.2006.282037
39. H. Chen, W. Ser, Sound source DOA estimation and localization in noisy reverberant environments using least-squares support vector machines. J. Signal Process. Syst. **63**, 287–300 (2011). https://doi.org/10.1007/s11265-009-0423-7
40. J. Nikunen, T. Virtanen, Direction of arrival based spatial covariance model for blind sound source separation. IEEE/ACM Trans. Audio Speech Lang. Process. **22**, 727–739 (2014). https://doi.org/10.1109/TASLP.2014.2303576
41. J.A. Belloch et al., Practical considerations for acoustic source localization in the IoT era: platforms, energy efficiency, and performance. IEEE Internet Things J. **6**, 5068–5079 (2019). https://doi.org/10.1109/JIOT.2019.2895742
42. K. Hoshiba, I. Komatsuzaki, N. Iwatsuki, Proposal of practical sound source localization method using histogram and frequency information of spatial spectrum for drone audition. Drones **8**, 159 (2024). https://doi.org/10.3390/drones8040159
43. H. Gulzar et al., FPGA based power-efficient edge server to accelerate speech interface for socially assistive robotics, in *2023 IEEE/SICE International Symposium on System Integration (SII)* (2023), pp. 1–6. https://doi.org/10.1109/SII55687.2023.10039093
44. R. Barona, E.A. Mary Anita, A survey on data breach challenges in cloud computing security: issues and threats, in *2017 International Conference on Circuit, Power and Computing Technologies (ICCPCT)* (Kollam, 2017), pp. 1–8. https://doi.org/10.1109/ICCPCT.2017.8074287
45. S. Branco, A. Ferreira, J. Cabral, Machine learning in resource-scarce embedded systems, FPGAs, and end-devices: a survey. Electronics **8**(11), 1289 (2019). https://doi.org/10.3390/electronics8111289
46. Md. Maruf Hossain Shuvo et al., Efficient acceleration of deep learning inference on resource-constrained edge devices: a review. Proc. IEEE **11**(1), 42–91 (2023). https://doi.org/10.1109/JPROC.2022.3226481
47. Z. Wang et al., Shuhai: benchmarking high bandwidth memory on FPGAS, in *2020 IEEE 28th Annual International Symposium on Field-Programmable Custom Computing Machines (FCCM)* (Fayetteville, 2020), pp. 111–119. https://doi.org/10.1109/FCCM48280.2020.00024
48. R. Schmidt, Multiple emitter location and signal parameter estimation. IEEE Trans. Antennas Propagat. **34**, 276–280 (1986). https://doi.org/10.1109/TAP.1986.1143830
49. H. Nakajima et al., Adaptive step-size parameter control for real-world blind source separation, in *Proceedings of the 2008 IEEE International Conference on Acoustics, Speech and Signal Processing* (Las Vegas, 2008), pp. 149–152. https://doi.org/10.1109/ICASSP.2008.4517568
50. Z. Hou et al., An FPGA off-loading of HARK sound source localization, in *Proceedings of the 2022 Tenth International Symposium on Computing and Networking Workshops (CANDARW)* (Himeji, 2022), pp. 236–240. https://doi.org/10.1109/CANDARW57323.2022.00057
51. Z. Qin et al., Low power implementation of Geometric High-order Decorrelation-based Source Separation on an FPGA board, in *Proceedings of the 2023 IEEE Symposium in Low-Power and High-Speed Chips (COOL CHIPS)* (Tokyo, 2023), pp. 1–6. https://doi.org/10.1109/COOLCHIPS57690.2023.10121954
52. Z. Lin et al., FPGA-based low power acceleration of HARK sound source localization, in *2024 IEEE Symposium in Low-Power and High-Speed Chips (COOL CHIPS)* (2024), pp. 1–6. https://doi.org/10.1109/COOLCHIPS61292.2024.10531180

Chapter 6
New Devices

Morihiro Kuga, Masahiro Iida, and Hideharu Amano

1 SLMLET

In recent years, there have been attempts to equip 5G relay stations with a certain degree of computing power, such as Multi-access Edge Computing (MEC) [1], to replace some of the computational processing currently performed in the cloud. This method is also excellent in that it avoids the risk of transferring important information to the cloud in terms of security and privacy and allows timing-critical processing to be performed without worrying about the cloud or network load. To perform such processing, instead of conventional simple IoT devices, high-performance computing power is required to support AI recognition processing while at the same time low cost and low power consumption. Recent FPGAs, such as SoC types [2] equipped with a CPU on the chip, have become common and can play a major role in MEC in terms of functionality and performance, but in terms of cost and power consumption, they are far inferior to dedicated ASICs. On the other hand, ASIC design, even if legacy processes are used for IoT, has high development costs and low flexibility, making it difficult to meet diverse requirements. The Shell–Role architecture has a CPU, memory, and general I/O in the Shell section, and a standard

M. Kuga (✉)
Faculty of Advanced Science and Technology, Kumamoto University, Kumamoto, Japan
e-mail: kuga@cs.kumamoto-u.ac.jp

M. Iida
Research and Education Institute for Semiconductors and Informatics, Kumamoto University, Kumamoto, Japan
e-mail: iida@cs.kumamoto-u.ac.jp

H. Amano
Systems Design Laboratory, Graduate School of Engineering, Bunkyo-ku, Tokyo, Japan
e-mail: hunga@dlab.t.u-tokyo.ac.jp

H. Amano (ed.), *Computation Platforms for Multi-access Edge Computing*,
https://doi.org/10.1007/978-981-96-8935-4_6

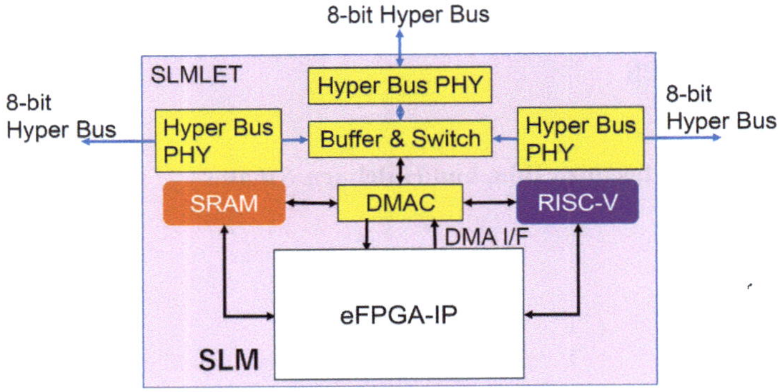

Fig. 6.1 A design concept of SLMLET

interface between the Shell section and the Role section, which significantly reduces design costs. On the other hand, the Role section can accommodate a variety of applications by including user-defined purpose-specific circuits. SLMLET has an area-efficient embedded FPGA in the Role section, making it user-programmable. Figure 6.1 shows a design concept of SLMLET with some functional blocks. An overview of each module is provided below, with further details available in Reference [3].

RISC-V Processor Core: RISC-V is a royalty-free, open-hardware architecture. You can choose from RISC-V products from around the world. For SLMLET, RISC-V is sufficient with a simple pipeline implementation using the RV32I instruction set. The core has memory blocks of appropriate size for instructions and data. To start the system, program binaries are loaded into the instruction and data memories via a serial peripheral interface (SPI) or universal asynchronous receiver transceiver (UART).

SLM-Based eFPGA-IP: An embedded FPGA-IP is integrated into the SLMLET chip. This eFPGA-IP has original logic blocks named Scalable Logic Cell (SLM) [4]. SLM can reduce the amount of configuration memory compared with the same input LUT and logic size. Section 2 provides a detailed explanation of the eFPGA-IP and the SLM architecture.

The connection between the eFPGA-IP and the RISC-V core gives a memory-mapped interface and is utilized for control and status data transfer. Two banks of shared memory are connected to the right and left sides of eFPGA-IP through the interconnect, respectively. In addition, Direct Memory Access Controllers (DMACs) offer the capability of streaming data transfer between the eFPGA-IP and the HyperBus. The configuration controller transfers the configuration data (i.e., bitstream of the FPGA) sets from the shared memory to eFPGA-IP. This action is triggered by the RISC-V software, and a dedicated controller autonomously controls the configuration itself.

Peripheral Interfaces: Assuming that the SLMLET is used as an edge device, the SPI is sometimes insufficient to handle data to/from the network. In addition, data-intensive applications, which require larger memory space than on-chip memory capacity, are also expected to be emerging IoT applications. Therefore, the SLMLET has three channels of HyperBus interfaces. HyperBus is developed by Cypress Semiconductor and is in accordance with JEDEC eXpanded Serial Peripheral Interface (xSPI) standard. Despite the reduced number of pins, it achieves high-speed data transfer, and commercially available HyperRAM and HyperFlash can be connected to the interface. Notably, HyperRAM offers lower power consumption compared to conventional DDR memory, making it suitable for edge devices. Additionally, the HyperBus ports can be connected to other SLMLET chips to extend the performance with parallel processing.

2 SLM-Based eFPGA-IP

2.1 Architecture Overview

An eFPGA-IP structure is shown in Fig. 6.2 by repeating the logic TILEs on the chip in an array style. In our design, all the logic tiles have a homogeneous structure to facilitate testing [5]. The optimal array size (x, y) can be found for the target

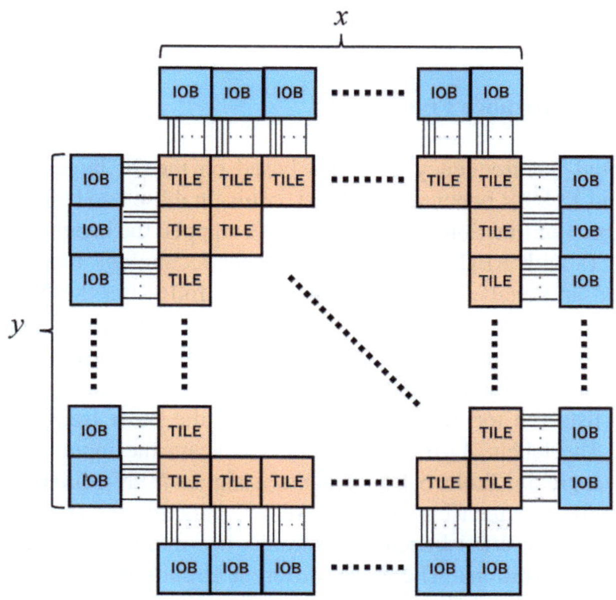

Fig. 6.2 Tile-based eFPGA-IP overview

application domain during architecture exploration. I/O blocks (IOBs) surround the outside of the tile array. An IOB is responsible for transferring a signal between a routing channel and an I/O cell outside the FPGA-IP. The homogeneous structure of the FPGA array requires only two tiles, logic tile and IOB tile. This also reduces the layout design time because only two layout patterns are required when designing the layout of tiles.

We designed and implemented a prototype chip called eFPGA-IP (TEG1), the details of which are described in the next section [6].

2.2 *Scalable Logic Cell (SLM)*

2.2.1 Why Do We Propose SLM?

Most modern FPGAs use an LUT as their basic logic cell. The circuit size of the LUT increases as $O(2^k)$ with an increasing number of inputs, k, so LUTs with more than six inputs negatively affect the overall FPGA performance. There are some logic cells derived from the subset of frequently used logics [7–9]. These logic cells implement only partial logics to optimize circuit area. However, the logic appearance frequency of different types of circuits changes, and it is impossible to use one function-reduced logic cell to express all circuits efficiently. In this work, we propose a novel SLM as a logic cell for FPGAs. The SLM shrinks the logic cell by also reducing logic express capability, but an SoC designer can explore the optimal SLM from a set of SLM structures for the target application to find an efficient domain-specific SLM logic cell for the FPGA-IP.

2.2.2 Logic Appearance Frequency

We investigated the frequency of logics based on the 20 largest MCNC benchmarks [10] and 13 VPR benchmarks [11]. Circuits in these two benchmarks are very different in terms of functionality and scale. We mapped the benchmark circuits into five-input LUTs with ABC and then analyzed the technology mapping results. We then classified and analyzed the mapping results following the NPN-equivalence class method [12], as in our previous work [9]. The NPN-equivalence class of a function consists of functions that can be derived by performing input negation, input permutation, or output negation on the original function.

Table 6.1 lists the top ten logics that appear the most frequently in the two benchmarks. The labels A, B, C, D, and E indicate the inputs of the five-input logics. Each logic in Table 6.1 represents a single NPN-equivalence class. We find that the most frequent logics in the MCNC benchmarks are simple AND-OR operations. However, VPR benchmark circuits use more logics, such as majority voter and XOR operations. Therefore, different circuits may have different high-frequency logics, so an application-specific logic cell made from one benchmark set may be inefficient for other application domains.

Table 6.1 Logic appearance rank (five inputs)

Rank	MCNC	VPR
1	$A \cdot B \cdot C \cdot D \cdot E$	$A \cdot \alpha + A \cdot B + B \cdot \alpha$
2	$A \cdot (B + C) \cdot (D + E)$	$A \oplus B \oplus \alpha$
3	$A \cdot B \cdot C \cdot (D + E)$	$A \oplus B \oplus C \oplus D \oplus E$
4	$A \cdot E + B \cdot C \cdot D$	$A \cdot B \cdot +(A + B) \cdot (C \cdot D \cdot E + (C \oplus D \oplus E))$
5	$A \cdot (B + C + D + E)$	$A \oplus B \oplus C \oplus D + \bar{A} \cdot C \cdot D \cdot E + A \cdot \bar{C} \cdot \bar{D} \cdot \bar{E}$
6	$A \cdot B \cdot (C + D \cdot E)$	$A \oplus B \oplus C \oplus (D \cdot E)$
7	$A \cdot (B + C + D \cdot E)$	$A \cdot (\bar{C} + \bar{D} + \bar{E}) + B \cdot C \cdot D \cdot E$
8	$A \cdot (B \cdot (\bar{D} + \bar{E}) + C \cdot D \cdot E)$	$A \oplus ((B \cdot E) \oplus (C \cdot D))$
9	$A \cdot B \cdot (C + D + E)$	$A \cdot B \cdot C + (A + B \cdot C) \cdot (D \oplus E)$
10	$A \cdot (B + C \cdot (D + E))$	$A \cdot (B + C) \cdot (D + E)$
		$\alpha = C \cdot D + D \cdot E + C \cdot E$

We then analyzed frequently appearing logics by performing the Shannon expansion to find common characteristics of these functions. Shannon expansion is one of the most basic logic conversion methods and is defined as

$$\begin{aligned} &f(x_0, x_1, \cdots, x_{k-1}, x_k, x_{k+1}, \cdots, x_{n-1}) = \\ &x_k \cdot f(x_0, x_1, \cdots, x_{k-1}, 1, x_{k+1}, \cdots, x_{n-1}) \\ &+\overline{x_k} \cdot f(x_0, x_1, \cdots, x_{k-1}, 0, x_{k+1}, \cdots, x_{n-1}). \end{aligned} \tag{6.1}$$

The two derived functions are called partial functions of the original function f. If we perform the Shannon expansion on input argument x_k of function f, it follows that

$$f = \overline{x_k} \cdot f_{(x_k=0)} + x_k \cdot f_{(x_k=1)}. \tag{6.2}$$

In addition, similar to the NPN-equivalence theory, partial functions $f_{(x_k=0)}$ and $f_{(x_k=1)}$ can be interconverted by negating or fixing the inputs or output. For example, a five-input logic function is

$$\begin{aligned} f = &A \cdot (C \cdot D + D \cdot E + C \cdot E) + A \cdot B \\ &+ B \cdot (C \cdot D + D \cdot E + C \cdot E). \end{aligned} \tag{6.3}$$

If we perform the Shannon expansion on argument A, partial functions $f_{(A=0)}$ and $f_{(A=1)}$ become

$$\begin{aligned} f_{(A=0)} &= B \cdot (C \cdot D + D \cdot E + C \cdot E), \\ f_{(A=1)} &= B + C \cdot D + D \cdot E + C \cdot E. \end{aligned} \tag{6.4}$$

Table 6.2 Ratio of interconvertible partial functions

	Item	5-LUT	6-LUT	7-LUT	8-LUT
MCNC	Ratio of k	57%	53%	45%	37%
	Coverage of k	98.34%	99.46%	99.53%	91.33%
VPR	Ratio of k	32%	28%	32%	25%
	Coverage of k	97.31%	87.18%	95.08%	81.46%

We then negate arguments $B,\ C,\ D,\ E$ and the output of $f_{A=0}$, obtaining

$$\begin{aligned}
&\overline{f_{(A=0, B=\overline{B}, C=\overline{C}, D=\overline{D}, E=\overline{E})}} \\
&= \overline{\overline{B} \cdot (\overline{C} \cdot \overline{D} + \overline{D} \cdot \overline{E} + \overline{C} \cdot \overline{E})} \\
&= B + (C + D) \cdot (D + E) \cdot (C + E) \\
&= f_{(A=1)}.
\end{aligned} \tag{6.5}$$

In this example, the number of inputs of a small LUT core is $k - 1 = 4$ because $f_{(A=0)}$ has four inputs, and the number of negated inputs is $w = 4$ because arguments $B,\ C,\ D,\ E$ are inverted.

Table 6.2 gives the ratio of interconvertible partial functions, where "Ratio of k" means the ratio of k input logic functions to all logic functions, and "Coverage of k" means the percentage of interconverted k input logic functions to all k input logic functions. For example, for five-input LUT mapping for MCNC, the ratio of five-input and less than five-input logic functions is 57% and 43%, respectively. 98.34% of five-input logic functions can be interconverted. The results show that partial functions of most logics in the two benchmark sets can be interconverted.

2.2.3 SLM Structure

We can obtain two partial functions by performing the Shannon expansion on any logic, and one of the partial functions will have a large potential for conversion to the other by only negating or fixing its I/Os while retaining core functionality. This means that we can reduce the configuration memory of a conventional LUT while still expressing the same logic. So, we can propose an SLM that can implement logic functions with smaller area and less configuration memory than those of an LUT.

Figure 6.3a shows the general structure of the SLM, which has three main parts: an input controller, an LUT core, and an output controller. For a partial function with interconvertible logic, the LUT core implements the shared core function of its partial functions, and the input and output controllers switch between partial functions. Figure 6.3b shows the circuit and the truth table of the programmable NAND. This circuit can negate inputs or fix output to 0 or 1 with two configuration memory cells.

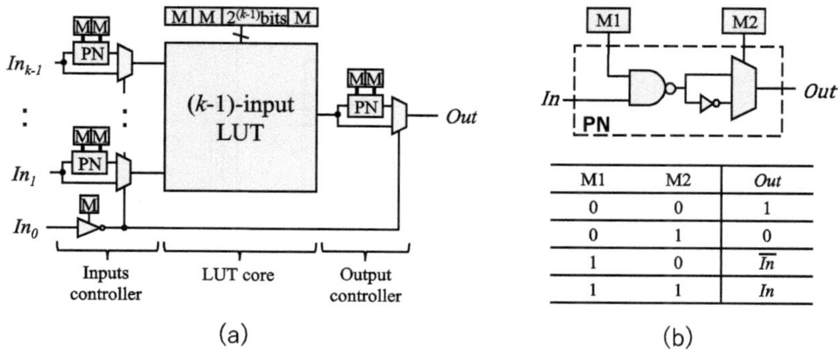

Fig. 6.3 SLM logic cell structure. (**a**) SLM structure. (**b**) Programmable NAND

Table 6.3 Available logic cells in Architecture Library

Logic cell	# of mem.	Area [um^2]	Delay [ps]
4-LUT	17	198.00	327.75
5-LUT	33	370.59	349.34
5-SLM (4,0)	20	235.87	473.00
5-SLM (4,1)	22	261.27	632.45
5-SLM (4,2)	24	286.68	654.56
5-SLM (4,3)	26	312.08	667.65
5-SLM (4,4)	28	337.48	682.53
6-LUT	65	725.99	371.89
6-SLM (5,0)	36	408.47	484.68
6-SLM (5,1)	38	433.87	644.85
6-SLM (5,2)	40	459.27	665.68
6-SLM (5,3)	42	484.67	678.79
6-SLM (5,4)	44	510.07	692.32
6-SLM (5,5)	46	535.47	706.92

When designing the input controller, it is unnecessary to make all inputs controllable for most partial function conversions. This allows trade-offs between SLM logic coverage and area to find the most area-efficient SLM by reducing the number of input controllers. The general k-SLM structure has k inputs (w of which are controllable, with $w \leq k-1$), one output, and a $(k-1)$-LUT core, which is designated as k-SLM $(k-1, w)$. For some logics, we need to convert $f_{(x=0)}$ to $f_{(x=1)}$, whereas other logics require the reverse conversion. The output controller is programmable for these two cases.

Table 6.3 gives the number of configuration memories, area, and delay comparisons of LUTs and SLMs with four to six inputs. LUTs are listed for comparison. The area and delay are derived by synthesizing these logic cells with the

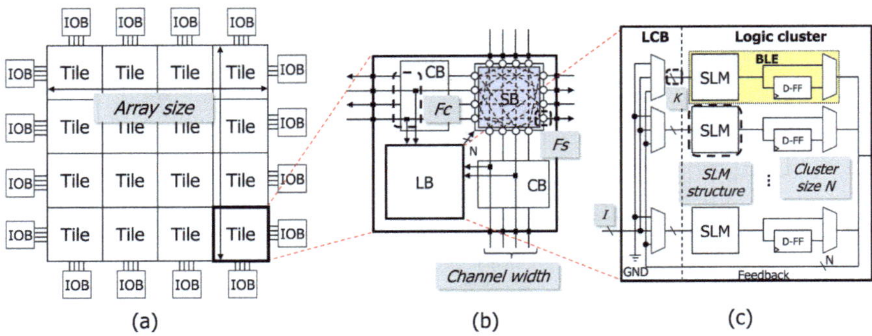

Fig. 6.4 eFPGA-IP architecture. (**a**) FPGA core. (**b**) Tile. (**c**) Logic block (LB)

USJC[1] 55nm LVT standard cell process. We can see that all SLMs have a smaller memory usage and area than LUTs with the same number of inputs. However, the delay of SLMs is longer because of the additional delay of I/O controllers. The logic cells listed in Table 6.3 are all available in our Architecture Library, and we can compare their performances when implementing the target application to find an optimal cell for the FPGA-IP.

2.3 eFPGA-IP Architecture

Figure 6.4 shows the (a) FPGA core, (b) tile, and (c) logic block designs of the proposed FPGA-IP. The gray labels indicate architecture parameters that can be optimized for domain-specific application circuits. The details of each part are explained below.

2.3.1 FPGA Core

An FPGA top design is constructed as shown in Fig. 6.4a by repeating the logic tiles on the chip in an array style. In our design, all the logic tiles have a homogeneous structure to facilitate testing [5]. The optimal *array size* can be found for the target application domain during architecture exploration.

On the outer sides of the tile array, there are $4 \times array\ size$ I/O blocks (IOBs). An IOB is responsible for transferring a signal between a routing channel and an I/O cell outside the FPGA-IP. Normally, FPGA I/O cells for user pins are bidirectional, but the routing wires are unidirectional, so an IOB can be configured as an input or an output and to map a selected wire in a routing channel to the I/O.

[1] United Semiconductor Japan Co., Ltd.

2.3.2 Tile

As Fig. 6.4b shows, a tile consists of an LB and global routing resources, the latter being used to connect the fine-grained logics distributed in many LBs to implement the user circuit. The global routing resources include two routing channels, two connection blocks (CBs) named CBT and CBR, and a switch box (SB). Each routing channel contains a group of routing wires. The parameter $channel\ width$ indicates the quantity of wires within a routing channel; each wire has only one direction for signal transfer. The CBs are responsible for connecting the routing channels to the LB. The parameter Fc indicates the ratio of wires connected to the LB. In the case of Fig. 6.4b, we have $channel\ width = 4$ and $Fc = 0.5$, so two out of four wires in a routing channel are connected to the LB. The SB is a switch of global routing channels, and we adopt the Wilton-type SB [13] whose structure is shown in Fig. 6.4b. The Wilton-type SB has $Fs = 3$, which means that any output multiplexer of the SB can switch its value from three other input directions. In our design, all LB outputs are connected to global routing through SBs, so an SB output multiplexer has $3 + N$ inputs. The optimal Fc and *channel width* can be found for the target application domain during architecture exploration.

2.3.3 Logic Block

As in Fig. 6.4c, a basic logic element (BLE) consists of an SLM with a scan FF and implements the most fine-grained logic and supports both combination and sequential modes: When a BLE works in combination mode, the SLM output is output directly; when a BLE works in a sequential mode, the SLM output is first stored in the FF and then output from there at the next clock cycle. The architecture parameter for the BLE is the SLM structure. The parameter K indicates the number of inputs of the SLM.

Then N BLEs are grouped into a logic cluster, where N indicates the cluster size. The BLEs within a logic cluster can be connected with low-latency local connections, thereby reducing the burden of global routing resources. A local connection block (LCB) is used to forward outside inputs to BLE inputs, as well as feedbacks of BLE outputs to other BLE inputs within the same cluster. The logic cluster and the LCB form a logic block (LB), as shown in Fig. 6.4a. The optimal *SLM structure* and the value of *cluster size* can be found for the target application domain during architecture exploration. To save the global routing resources, the number of inputs of an LB, I, is set as less than the number of inputs of all BLEs ($K \times N$), and I is calculated as [14]

$$I = \frac{K}{2} \times (N + 1). \tag{6.6}$$

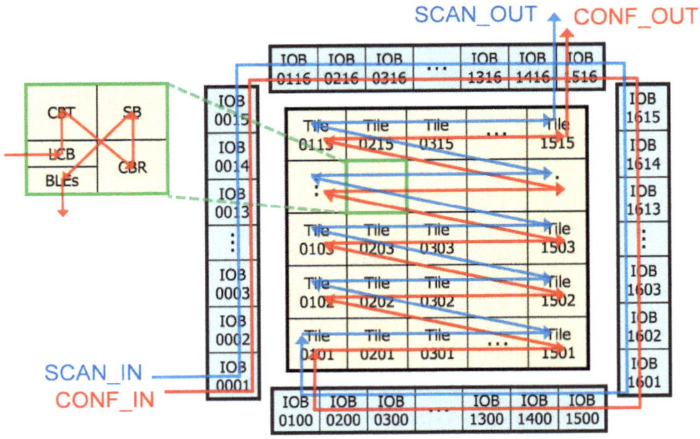

Fig. 6.5 Configuration chain and scan chain

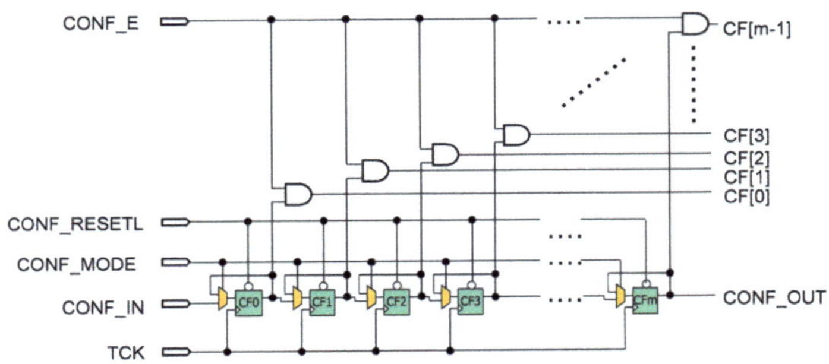

CONF_MODE	Function
1	Configuration mode (Shift operation)
0	Execution mode (Hold operation)

Fig. 6.6 Configuration circuit

2.3.4 Configuration Chain and Scan Chain

In our design, all configuration FFs (CFs) are connected to form a configuration chain, which is shown by the red arrows in Fig. 6.5. Figure 6.6 shows the circuit of the configuration chain. Along this path, a single FF stores one configuration datum. A bitstream of an FPGA consists of all configuration bits arranged in the same order as that of the configuration chain. During configuration, we first disable the CONF_E signal so that all configuration bits in the FPGA receive a stable 0. We then enable the CONF_MODE signal to allow CFs to read values from their inputs. Then the bitstream is input from the CONF_IN port. The configuration is performed by shifting the bitstream according to the clock TCK. When all configuration bits

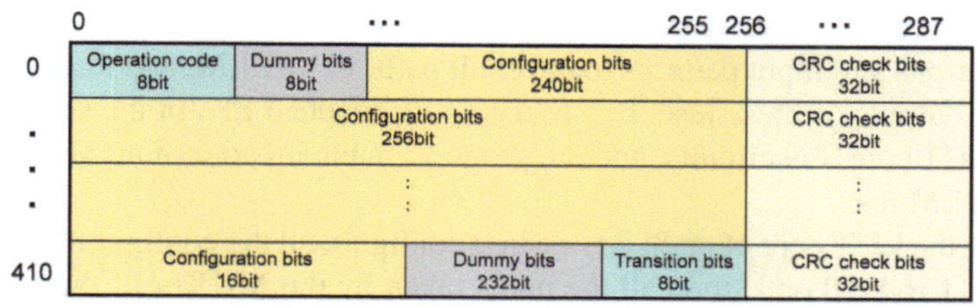

Fig. 6.7 Bitstream format

are shifted to their CFs, we disable CONF_MODE and enable CONF_E to transfer the FPGA status from the configuration mode to the user mode.

In addition, there is a scan chain to read out the values of all FFs of BLEs. The scan path is shown by the blue arrow in Fig. 6.5. A scan chain consists of a series of scan FFs that store the results of the user circuit or chip tests.

2.3.5 Controller

Besides the FPGA, we implement a controller to manage the FPGA-IP status, i.e., configuration mode, user mode, and test mode. One of the main functions of the controller is to decode and execute the bitstream. Figure 6.7 shows an example of the format of a bitstream with 104,960 configuration bits. A bitstream consists of multiple 288-bit-long frames. Each frame carries 256-bit configuration bits and 32-bit cyclic redundancy check code. At the final frame, if there are fewer than 256 configuration bits, the rest are filled as dummy bits. The first 8 bits of a bitstream are the instruction code that sets up the mode of the FPGA; the transition bits of the final frame are the transition code that specifies the next state to transfer to after this bitstream.

2.3.6 Shipping Test

In general LSI, the automatic test pattern generator (ATPG) tool is used for testing. However, in FPGAs, the test pattern cannot be generated by the ATPG tool because the user circuit is not uniquely determined. In addition, the routing part of an FPGA involves a huge number of switch combinations, which makes it difficult to implement the test circuit. In our eFPGA suite, we adopt a previously developed FPGA test method in reference [15] for stuck-at faults in logics and routing resources.

We choose long line for device delay evaluation of eFPGAs. Here, long line is defined as a path that traverses the FPGA from top to bottom, left to right, and right to left. In discrete FPGAs, a longline is often defined as a path that traverses FPGA without switches, but note that here it passes through the routing part. The evaluation of the long line delay will be discussed in Sect. 3.7.2.

For the test of an SLM shown in Fig. 6.3a, we use test patterns with different configurations and input patterns to cover all paths of an SLM, and then we examine the SLM outputs of these tests by observing its attached FFs in the same BLE. We test the LUT core, PN circuits, and the programmable inverter of an SLM to achieve a 100% SLM test.

To test the LUT core of an SLM, we first configure all the configuration memories of the LUT to be 0 or 1; then all the paths taken by the MUXes of the LUT can be activated by testing all input patterns of the LUT. For the PN circuits of the SLM, we configure all the configuration memories of the LUTs to 0 or 1, and we test all the PN patterns shown in Fig. 6.3b; then we can test for 0/1 stuck-at faults in each path of the PN circuits. For the test of the programmable inverter on the In_0 input path, the output PN circuit is configured as inverted output; then the input In_0 works as a switch of the SLM to be either the LUT output or the inverted LUT output.

3 A Case Study of Domain-Specific eFPGA-IP (TEG1)

In this section, we show the FPGA-IP architecture exploration, chip design, shipping test, and performance evaluation for an edge-oriented DP application circuit as a case study.

3.1 Target Application

DP is an algorithm that protects the privacy of individuals during data collection and analysis. Because low-end microprocessors on edge devices usually cannot perform as required for DP computation, we must implement hardware acceleration to improve performance. In addition, the DP circuit requires the algorithm to be upgraded for security enhancement at runtime, so it is suitable to take the implementation approach of embedding an eFPGA into the SoC and then implementing the DP circuit on the eFPGA, which can provide sufficient performance to perform real-time DP processing of data collected by the edge device. Our target DP circuit has a data width of 8 bits. The DP circuit processes input data with random values, so that the exact value of the input cannot be identified from the output data, but the statistical characteristics of the original data are kept for analysis.

3.2 Architecture Exploration

We performed the FPGA-IP architecture exploration for the target DP circuit based on the flow shown in Fig. 6.11(1). The architecture parameters to be determined during exploration are shown as gray labels in Fig. 6.4. First, we chose all SLM structures with five or six inputs as logic cell candidates, and we evaluated cluster sizes of 4 and 8. The exploration results are shown in Fig. 6.8. From (a) and (b), we can see that the combination of 5-SLM(4,0) with a cluster size of four has the smallest configuration memory usage and total area. For the critical path delay, (c) shows that 6-SLM(5,1) with a cluster size of 4 performs the best. On the metric of the area–delay product, 6-SLM(5,1) with a cluster size of 4 has the best overall performance. So we can see that the choice of SLM structure and cluster size will differ according to the design target. In this work, we chose the smallest 5-SLM(4,0) with a cluster size of 4 to minimize the cost of silicon area on a chip.

Based on the logic exploration results, other parameters are then optimized for the DP circuit, and the specifications of the FPGA-IP are given in Table 6.4. The channel width was set to 30, which is the minimum channel width for a router to successfully complete the DP circuit routing. In this work, we set the array size to 16×16 to fully fill the given 4.41 mm^2 chip area. In an SoC design, we can set the minimum array size to provide the number of tiles required by the target circuit to save chip area usage. The DP circuit implementation uses 158 out of a total of 256 CLBs (62%). For the bitstream, 104,960 bits are used as FPGA configuration memory; there are additional bits for instruction and error handling, so the total bitstream size is 118,368 bits.

3.3 Chip Design

When the architecture parameters are fixed, we obtain the generated optimal eFPGA-IP design from the end of the flow shown in Fig. 6.11(1). We designed and fabricated the eFPGA-IP (TEG1) chip using the USJC 55 nm standard cell process. Figure 6.9 shows the (a) evaluation board, (b) chip top floor planning, and (c) tile layout. The tile size in Fig. 6.9c is 100.8 um square. We manually planned the positions of the macros of BLEs, LCB, CBs, and SB and then let configuration memories fill the remaining spaces. The SB is placed at the center to minimize the delay variation of different routing paths. From the chip in Fig. 6.9b, we can see that the main tile array occupies most of the area.

3.4 Evaluation Board

We have developed an evaluation board. The block diagram is shown in Fig. 6.10. It is designed as a daughter board of PYNQ-Z2 by Tul using a Zynq-7000 programmable SoC by Xilinx so that the control signal and the FPGA I/O of the chip can be flexibly controlled. In PYNQ-Z2, 59 I/Os are connected to the PL (Programmable Logic) part of the Zynq device. By building a control circuit in the PL part according to the purpose of evaluation, the chip can be easily controlled by the PYNQ environment from the ARM processor in the PS (Processing System) part. A

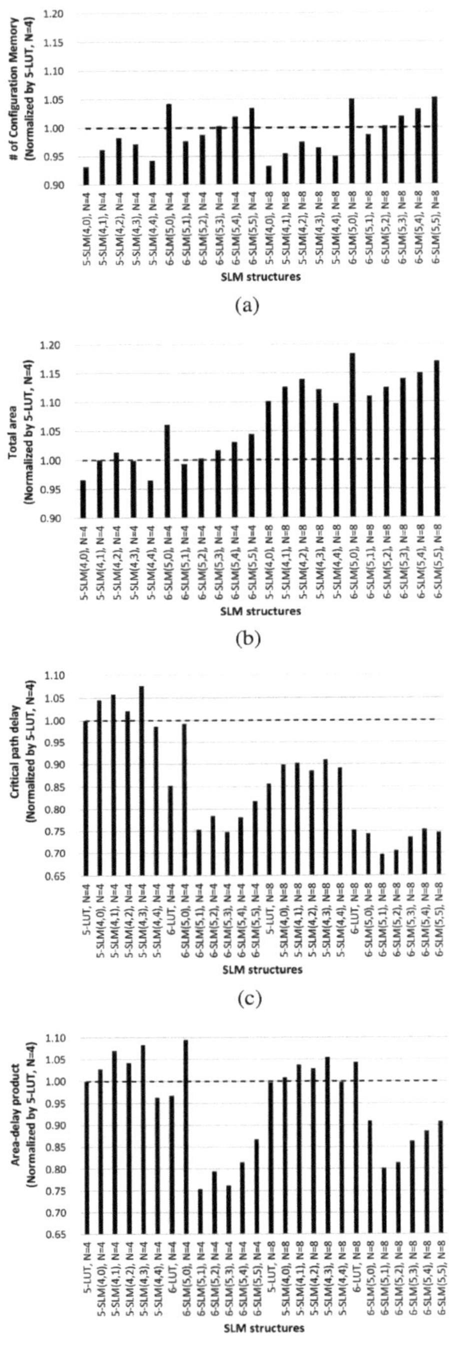

Fig. 6.8 SLM structure exploration results. (**a**) # of configuration memory. (**b**) Total area. (**c**) Critical path delay. (**d**) Area-delay product

Table 6.4 The eFPGA-IP (TEG1) chip specifications

Item	Value
Array size	16 × 16
Channel width	30
Logic cell	5-SLM(4,0)
Logic clusters	4
LB inputs	12
SB structure	Wilton (Fs = 3)
CB structure	normal (Fc = 0.25)
I/O pins	64
Conf. mem. (FPGA array)	104,960 bit
Bitstream size	118,368 bit
Die size	4.41 mm^2 (2.1 mm × 2.1 mm)
Process	USJC 55 nm LVT
Core voltage	0.9V
I/O voltage	1.8/2.5/3.3 V

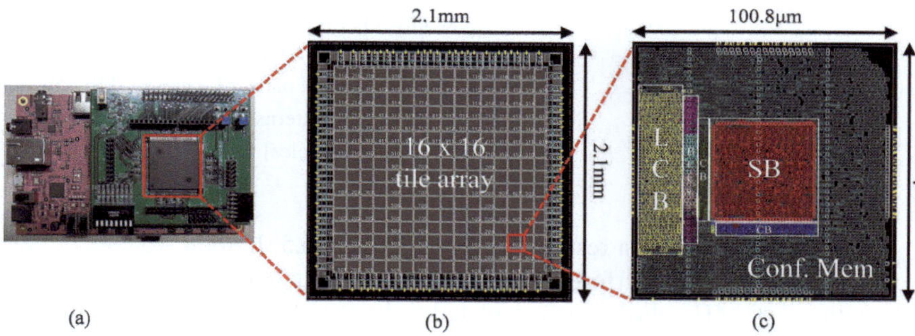

Fig. 6.9 Layout of the eFPGA-IP (TEG1) chip and its evaluation board. (**a**) Evaluation board. (**b**) Chip. (**c**) Tile

total of 59 signals, including 11 chip control signals and 48 out of 64 FPGA I/Os, can be connected to the PL part. The remaining 16 FPGA I/Os are connected to switches (SWs) and LEDs on the daughter board. The I/O voltage is set to 3.3 V to interface with the PYNQ-Z2, and the core voltage is set to 0.9 V for typical operation.

3.5 *Shipping Test*

To confirm the stuck-at fault testing coverage, we input test patterns to the FPGA and observed the toggle coverage using Synopsys VCS Q-2020.03-SP1. To test all hardware resources, we prepared a total of 128 test patterns, each of which was tested by repeating "configuration operation → normal operation → scan operation." The test results were first stored in the FFs in the BLEs and IOBs and then scanned out for analysis.

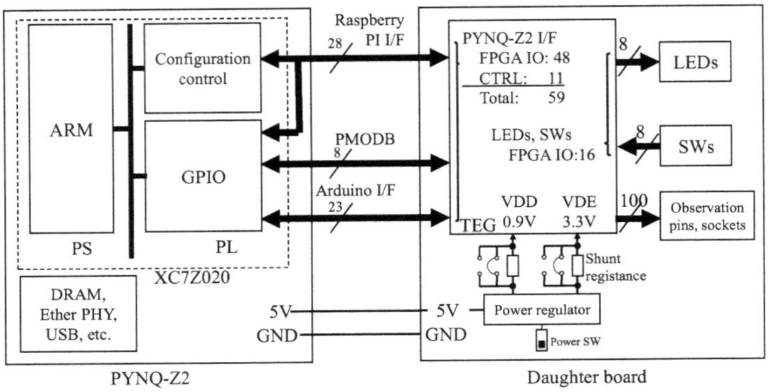

Fig. 6.10 Block diagram of the evaluation board

Table 6.5 Test cycle results

Item	Value
Normal operation [cycles]	2
Scan operation [cycles]	1087
Other control operations [cycles]	13,410
Test cycles per test pattern [cycles]	119,459
Number of test patterns	128
Total test time [cycles]	15,290,752

The test cycle evaluation results are given in Table 6.5. In total, 15,290,752 test cycles were required to perform the test with all 128 test patterns. The total test time with a 100-MHz clock was calculated to be 0.153 s. This indicates that our test method can complete the test in a short time for shipping test requirements. In addition, we observed 100% toggle coverage in the entire FPGA-IP using the proposed method, which means that 100% stuck-at faults can be detected during the shipping test.

3.6 eFPGA CAD Flow

Figure 6.11 shows an overview of the design flow of the proposed eFPGA generation suite. An FPGA-IP's architecture exploration, SoC implementation, and utilization are supported by three main flows: Exploration and generation of FPGA-IP architecture, chip implementation, and FPGA-IP utilization.

The FPGA-IP architecture exploration and generation flow explores the optimal FPGA architecture parameters for the application circuits provided by the SoC designer and then generates the Verilog of the optimal FPGA-IP and the bitstream for the shipping test. The SoC designer only needs to prepare domain-specific benchmark circuits and specify an architecture space for FPGA-IP exploration. The eFPGA Architecture Library contains all possible FPGA architectures supported

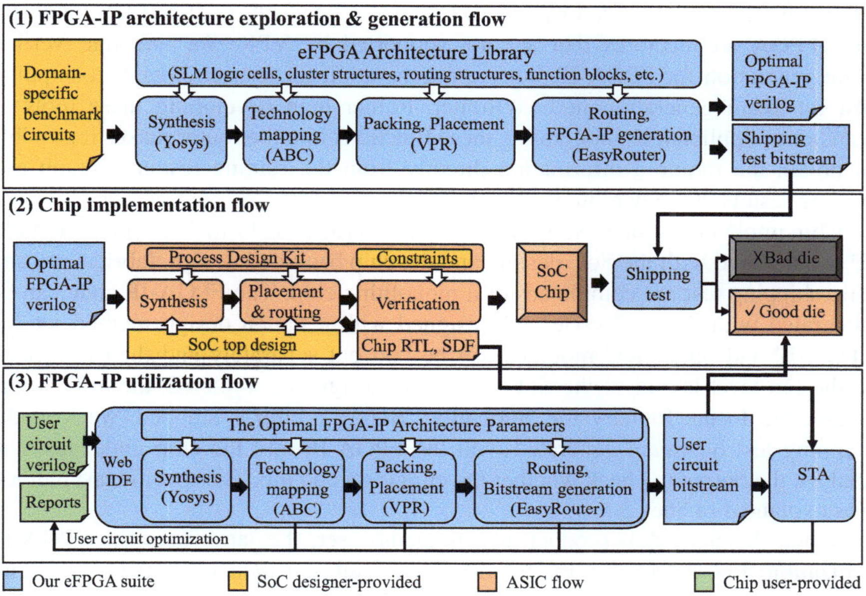

Fig. 6.11 Overview of design flow of proposed eFPGA suite

by our framework. An FPGA architecture is determined by a combination of architectural parameters that define the SLM structure, cluster structure, routing structure, function blocks, and other modules, as well as parameters that define the tile array structure. The architecture exploration flow implements benchmark circuits on each architecture selected by the SoC designer and gives a performance report, and then the SoC designer analyzes the performance report and finds the optimal FPGA-IP to be used based on the design goal (e.g., minimum area or shortest delay).

The CAD tool is used with the Verilog-to-Routing (VTR) [16] tool that is widely used around the world. However, some tools are modified to support the SLM logic block. Our CAD flow is explained below. First, the Verilog user circuit is synthesized with Yosys [17] into a Blif-format netlist. Second, we perform technology mapping on the netlist. The Architecture Library passes the information of an SLM structure to ABC [18]. To support SLM mapping, we added a Boolean matching function for the SLM to the FPGA mapping flow of ABC. In addition, we use an efficient function decomposition method to convert coarse-grained logics into fine-grained logics that can be implemented by the specified SLM. The details of the SLM will be introduced in Sect. 2.2. Then, we use VPR for packing and placement. The Architecture Library provides the standard VPR FPGA architecture description file to VPR, which contains all FPGA architecture information. Finally, we developed our own routing tool—the EasyRouter [19]—for FPGA routing and FPGA-IP generation. Developed in C#, EasyRouter supports

FPGA architectures defined by the VPR and can be easily modified to implement new FPGA architectures that are not supported by VPR. We integrate Verilog code generation for FPGA-IP, bitstream generation for user circuits, and shipping test bitstream generation in EasyRouter. Rather than developing these tools as standalone applications, integrating these functions into the router makes it possible to obtain the required information directly from the routing resource graph and routing results, thereby reducing the development effort.

Chip implementation flow is based on a standard ASIC flow. The optimal FPGA-IP is integrated with the SoC design, and the chip back-end design, fabrication, and chip shipping test are completed for a monolithic chip. The FPGA-IP code can be integrated with the SoC code and designed with a flatten design method, or the FPGA-IP and other SoC modules can be laid out separately and then integrated in the top-level layout of the SoC. After the design is completed, the SoC layout is tape out to the foundry for fabrication. Before a chip is shipped, we load the shipping test bitstream generated with the FPGA-IP into the chip under test and analyze the output to check whether the chip is a good die. The specific test method was explained in Sect. 3.5.

When the SoC chip product arrives at the user, the latter uses the FPGA-IP utilization flow to develop the application circuit and then loads the user-circuit bitstream into the FPGA module of the SoC to implement the circuit. Basically, the user-circuit design flow uses the same CAD tools as the FPGA-IP architecture exploration and generation flow, and all tools process the user circuit based on the optimal FPGA-IP architecture parameters. At this point, EasyRouter generates a bitstream of the user circuit after routing, and then the bitstream is loaded into the FPGA to complete the deployment of the circuit. We use OpenSTA [20] to perform accurate static timing analysis (STA) of the routed user-circuit bitstream with the SoC chip's register transfer level code and standard delay format (SDF) information extracted from the layout. Each CAD tool will return design reports to the user, who will know whether the compilation of the circuit is successful and whether the circuit design needs improvement based on these reports.

3.7 *Results and Discussion*

3.7.1 Delay Evaluation by STA

In this evaluation, we performed static timing analysis using Synopsys's PrimeTime P-2019.03-SP3 under the typical conditions, an operating voltage of 0.9V and the temperature of 25°C. First, we got the delay values of each path from the primary inputs to the primary outputs in the tiles of the FPGA. In the tested chip, there are 14,400 paths from input to output when the channel width is 30, there are four LB clusters, and the input and output have four directions. The DP circuit was implemented using the design flow described in Fig. 6.11(3).

Figure 6.12 shows the path delay distribution from tile inputs to tile outputs. The delays are distributed in the range of 1.43–1.98 ns, with an average value of 1.69 ns.

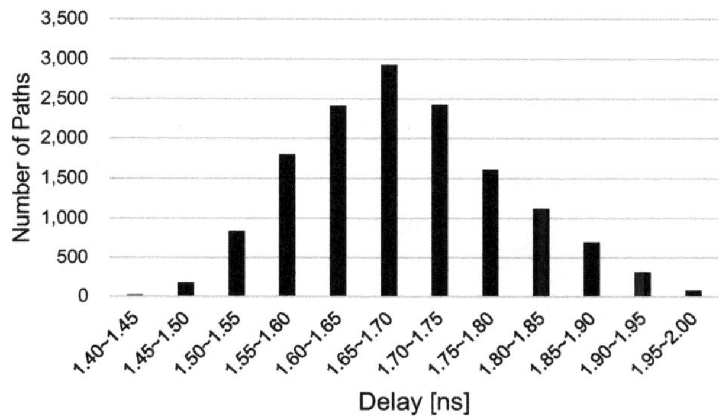

Fig. 6.12 Path delay distribution from tile inputs to tile outputs

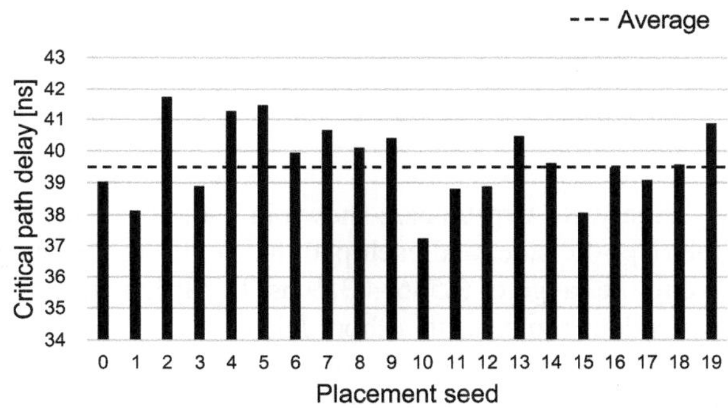

Fig. 6.13 Distribution of critical path delay for DP circuit implementation (20 initial seeds)

It can be seen that the delay variation is within about 6% of the mean value. The delay distribution of the target DP circuit is shown in Fig. 6.13. To examine the delay variation of the initial seed of placement, the initial seed was varied from 0 to 19 during placement, and the aeration of the critical path delay was checked. In this case, the delays are distributed in the range of 37.23–41.75 ns, with an average value of 39.68 ns. It can be seen that the delay variation is kept within 3%. So the DP circuit can operate at a maximum frequency of 26.8 MHz, and 1 byte of data can be processed at each clock cycle, so the throughput is 27.8 MOPS (Mega Operations Per Second). When Python executes DP processing, it is achieved about 2215 OPS (Operations Per Second). The DP circuit is 12.55×10^3 times faster than Python.

3.7.2 Long Line Delay on the Fabricated Chip

To evaluate long line delay on the fabricated chip, we use the test circuit shown in Fig. 6.14. To create a long line, connect the Q output of IOB-FF located on one side to the D input of IOB-FF located on the opposite side.

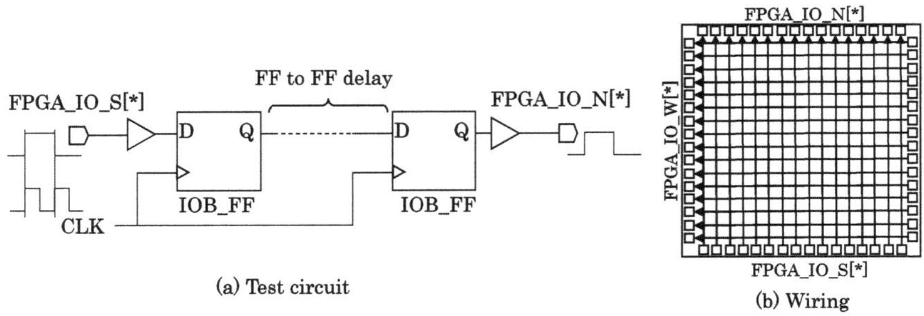

Fig. 6.14 Test circuit for long line delay

Table 6.6 STA results of long line delay

Path	Min. [ns]	Ave. [ns]	Max. [ns]
North to South	3.860	3.969	4.091
South to North	4.179	4.266	4.389
East to West	4.092	4.197	4.292
West to East	3.586	3.802	3.943

Firstly, STA results are shown in Table 6.6. There are 16 paths from one side to the opposite side, and each path has a different propagation delay. All path delays are in the range of 3.586–4.389 ns. In particular, the path from the west side to the east side is routed with a short delay.

Then, we observe the propagation status of the one-shot pulse when the frequency and core voltage are changed, and a Shmoo plot is created from the results. Figure 6.15 is one of the Shmoo plot results that indicates the operating status from the south side to the north side path. The Shmoo plot shows that the blue area indicates correctly operation, and on the other hand the red area indicates not. We confirm the correct behavior up to 250 MHz (4.0 ns) on the actual chip; however STA results show that the maximum delay is 4.389 ns, the minimum delay is 4.179 ns, and the average is 4.266 ns. It is confirmed that the operation is slightly faster than the result of STA.

3.7.3 Validation of an Application Behavior

To validate the application behavior, we implement two applications, a simple 24-bit binary counter and LDP (Local Differential Privacy) circuits. The implementation result is shown in Table 6.7. Both applications are correctly operated on the chip.

A 24-bit binary counter is implemented on the actual chip as a simple application. The STA result shows that the generated FPGA-IP can achieve a maximum frequency of 176.7 MHz (5.66 ns). However, the actual operating frequency is observed 117 MHz (8.55 ns). And also, the propagation delay of clock to pad is measured about 12 ns. Although the inside of the chip can operate at high speed, it should be noted that the delay of the I/O buffer is large. The power consumption is

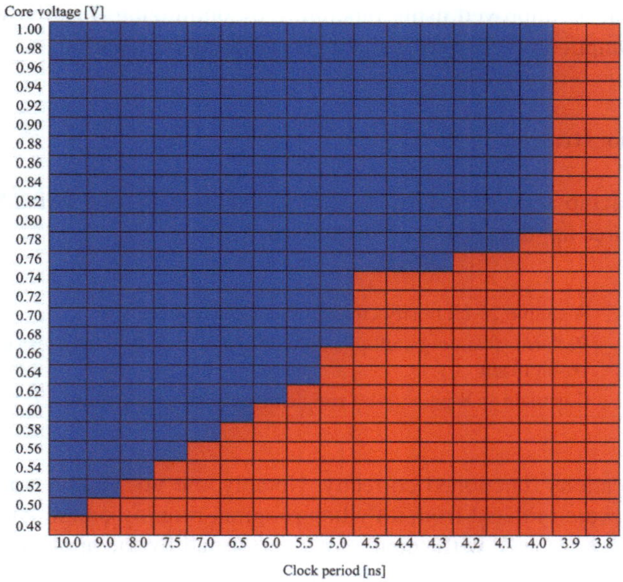

Fig. 6.15 Shmoo plot of long line delay

Table 6.7 Implementation result of application

Resource usage		
Module	24-bit counter	LDP
BLE (SLM)	39/1024 (3.8%)	993/1024 (97.0%)
LB	10/ 256 (3.9%)	250/ 256 (97.6%)
FF	24/1024 (23.4%)	102/1024 (10.0%)

Operating frequency and measured power		
Item	24-bit counter	LDP
Operating frequency [MHz]	117	10
Measured Power [mW]	162	139

measured about 162 mW under the condition of an operating frequency of 117 MHz and a core voltage of 0.90 V.

Next, we also implement an LDP (Local Differential Privacy) circuit to modify the interface of DP circuit used in Sect. 3.1. To connect a host processor on the evaluation board, the LDP has a streaming interface by AXI stream so that a lot of data is modified with random values using DMA (Direct Memory Access) transfer. And it is modified internal functions to reduce calculation error, so the circuit is larger than the DP circuit in Sect. 3.1. Resource usages of BLE and FF are 97.0% and 7.6%, respectively, so the LDP is a large application for the actual chip. The critical path delay is 57.3 ns (17.4 MHz) by STA. The delay is larger than the DP circuit in Sect. 3.1. It is a reason that the LDP circuit has a long critical path from random generator to the final adder for adding a noise. However the operating frequency can be improved using pipelining by optimizing. Finally, we measured power consumption of 139 mW at 10 MHz operating frequency.

3.7.4 Demonstration

During the demo session in FPL 2022, we demonstrated the eFPGA-IP (TEG1) chip [21]. Three applications were showcased:

`counter24`	A simple 24-bit binary counter with LED lighting
`vec16`	Perform streaming multiplication of two 8-bit packed binaries and output a 16-bit result
`sma8`	Calculate an 8-point, 8-bit simple mean average and return the result

All applications functioned correctly. Notably, the `vec16` and `sma8` streaming applications were initiated from a Python program running on the ARM processor of the PYNQ-Z2 board, with calculations performed by the application circuit on the eFPGA-IP (TEG1), and the results obtained.

3.7.5 In Comparison with Spartan-3 XC3S50

The developed eFPGA-IP (TEG1) has a logic resource scale roughly equivalent to the Xilinx's Spartan-3 XC3S50 device. The XC3S50 has a total of 192 Configurable Logic Blocks (CLBs) of 16 $\times$ 12 tiles. Each CLB has four slices, and each slice has two 4-LUTs, for a total of 1536 4-LUTs. Therefore, we conducted an implementation comparison of the DP circuit with the XC3S50.

The implementation result is shown in Table 6.8. The DP circuit was used as the application. It includes a 43-bit linear feedback shift register (LFSR) and a 37-bit cellular automata shift register (CASR) as a random number generator, a 16-bit multiplier, and an adder/subtractor. 80 FFs are used. As for the logic blocks, in the case of TEG1, a total of 634 SLMs are used. Of these, 195 are used for multipliers, 233 are used for circuits to calculate logarithmic values, and the rest are used for adders/subtractors.

For XC3S50, the results vary depending on the compile options used during logic synthesis using the Xilinx's ISE 14.7 tool. When using the MULT18x18 multiply-and-accumulate unit, the number of CLBs used can be significantly reduced. Also, since the logtable can use the LUT configuration memory as LUTRAM, the number of CLBs used can be reduced compared to TEG1. Even without using MULT18x18, the XC3S50 can implement a multiplier with only 80 CLUs. This is likely due to the fact that logic synthesis and technology mapping in ISE14.7 are functioning properly.

As described above, eFPGA-IP(TEG1) and XC3S50 have almost the same logic resources. However, when implementing a circuit on eFPGA-IP(TEG1), we found

Table 6.8 Logic resource usage, eFPGA-IP (TEG1) vs. XC3S30

FPGA		TEG1	Spartan-3 XC3S50		
Array size		16×16=256LBs	16×12=192CLBs (1CLB=4 Slice = 8LUTs)		
# BLEs		256×4=1024SLMs	192×8=1536LUTs		
Synthesizer		Yosys	ISE 14.7		Yosys+ISE14.7
Synth. opt.		non-flat	mult_style=auto	mult_style=LUT	
Resource	FF	SLM	LUT		
dp/	0/80	8/634	45/176	45/252	17/258
Multiplier	0/0	195/195	MULT18×18=1	77/77	80/80
add8_L_p	0/0	21/21	–	–	8/8
add8_datao	0/0	21/21	–	–	10/10
add8_L_m	0/0	21/21	–	–	8/8
sub9	0/0	22/22	–	–	Absorbed as add8
logtable	0/0	233/233	86/86	85/85	91/91
rand32	0/0	33/113	5/45	5/45	4/44
▻casr37	37/37	37/37	37/37	37/37	37/37
▻lfsr43	43/43	43/43	3/3	3/3	3/3
Total	80	634	176	252	258

that the following two points had a significant impact on performance: eFPGA-IP(TEG1) does not have multiply-accumulators and LUTRAMs, and its FPGA design tool is not optimized enough.

4 Improvement Toward the Next Versions of eFPGA-IP

After developing the eFPGA-IP (TEG1) chip, we made improvements for the next version and developed two versions. The eFPGA-IP (TEG2) was the second version for the SLMLET-1 chip, and the eFPGA-IP (TEG3) was the third version for the SLMLET-2.

4.1 The SLMLET-1 and the eFPGA-IP (TEG2)

4.1.1 The SLMLET-1

The eFPGA-IP (TEG1) was a prototype device with only the eFPGA-IP part implemented on the chip. Although the evaluation results showed that the FPGA functions worked properly, it was found that many issues remained to be improved. Therefore, we improved the eFPGA-IP (TEG1) and developed the second version, named eFPGA-IP (TEG2). This eFPGA-IP (TEG2) was implemented as the eFPGA-IP

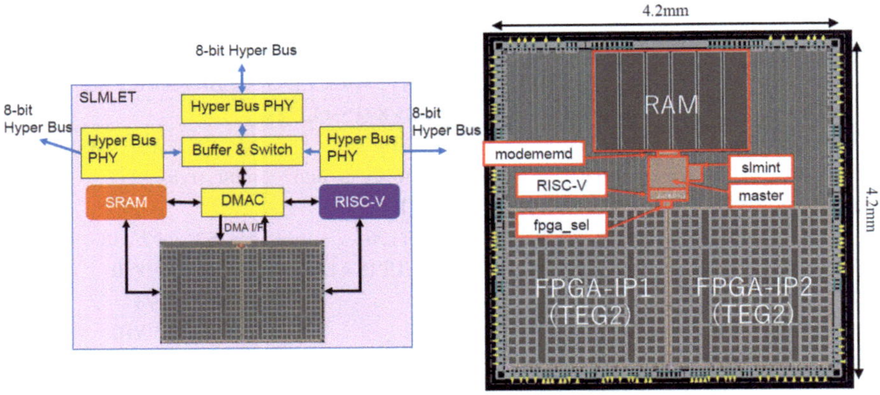

Fig. 6.16 Block diagram and layout image of the SLMLET-1 chip. (**a**) Block diagram. (**b**) Layout image

of the SLMLET-1 chip, along with a RISC-V processor, DMA, and SRAM [3]. Figure 6.16 shows the block diagram and the layout image of the SLMLET-1 chip.

The RISC-V processor is a three-stage pipeline implementation of the RV32I instruction set. The core has dedicated memory blocks, 64 kiB each, for instruction and data. When the system boots up, the program binary is loaded into the instruction memory and data memory via the Serial Peripheral Interface (SPI). The SLMLET can connect to up to four SPI slaves, including the boot loader.

Assuming that the SLMLET is used as an edge device, the SPI is sometimes insufficient to handle data to/from the network. In addition, data-intensive applications, which require larger memory space than on-chip memory capacity, are also expected to be emerging IoT applications. Therefore, the SLM- LET has three channels of HyperBus interfaces. HyperBus is developed by Cypress Semiconductor and is in accordance with the JEDEC eXpanded Serial Peripheral Interface (xSPI) standard. Despite the reduced number of pins, it achieves high-speed data transfer, and commercially available HyperRAM and HyperFlash can be connected to the interface. Notably, HyperRAM offers lower power consumption compared to conventional DDR memory, making it suitable for edge devices. Additionally, the HyperBus ports can be connected to other SLMLET chips to extend the performance with parallel processing.

4.1.2 The eFPGA-IP (TEG2)

The eFPGA-IP (TEG2) had a 16 × 16 array size, the same as the eFPGA-IP (TEG1), but had several improvements. For circuit resources, ADDER and DSPTILE are added. ADDER is a 1-bit full adder located in a BLE. When the BLE is in ADDER mode, the 4-LUT in the SLM is split into two 3-LUTs, and their outputs become the inputs of the ADDER. The four ADDERs in the four BLEs in a tile are connected as a 4-bit ripple carry adder (RCA), and the carries between tiles are cascaded. This allows the use of a high-speed carry chain without using the SLM, even though it is an RCA, which speeds up the adder/subtractor. DSPTILE is an 8-bit unsigned multiplier that calculates a 16-bit unsigned result. The height of DSPTILE is four tiles, and eight DSPTILEs are placed in an eFPGA-IP (TEG2).

4.1.3 Fabricated SLMLET-1 Chip

The SLMLET-1 chip was fabricated using the USCJ 55 nm DDC process. For details of the process, refer to [22]. Body bias control, which changes the bias voltage, can adjust the threshold voltage and reduce leakage power. It is possible to balance the trade-off between leakage power and computational performance. The RTL design of the SLMLET-1 was synthesized with a low voltage threshold (LVT) standard cell library using Synopsys's Design Compiler. The layout was designed with Cadence's Innovus. The photo of the fabricated chip is shown in the right figure of Fig. 6.16. The chip size is 4.2 mm square. The two square blocks at the bottom are two eFPGA-IPs (TEG2). The tile is laid out in a 106.80 um square area, and each tile is placed in a 16.20 um pitch. As Table 6.1 explains, the LB tile has four five-input SLM logic cells. Therefore, the density of the SLM logic cells is about 264 SLM/mm^2.

Several bugs were found in the fabricated SLMLET-1 chip, including some of the instruction operations of the RISC-V CPU, the interface between RISC-V and eFPGA-IP (TEG2), and the reversal of serial data transfer when performing hardware migration. It was possible to make the functions themselves work by changing the software and the hardware on eFPGA-IP (TEG2). As shown in reference [3, 23], some applications achieved performance improvements that exceeded those of current embedded CPUs. However, it was found that when the target application became large scale, it became difficult to implement it on eFPGA-IP (TEG2), resulting in a decrease in performance.

4.1.4 Issues for Improvement Identified from Evaluation Results

Through simulations and operational tests of the prototype SLMLET-1 chip, we have confirmed that there are problems with its functionality and issues to be addressed to improve its performance about eFPGA-IP (TEG2). In this subsubsection, we explain the problems and issues below.

4.1.4.1 Structural Problem

The eFPGA-IP (TEG1) was evaluated as fully operable as an eFPGA-IP, so when incorporating it into the SLMLET-1, eFPGA-IP (TEG2) was developed based on the eFPGA-IP (TEG1) architecture without changing the size of the array, and two of these were incorporated into an adjacent configuration. This made it possible to simultaneously execute different processes by implementing separate application circuits in each eFPGA and to execute large-scale processes by connecting the two and implementing a larger application circuit. In this case, the IOBs in the adjacent parts were wired so that they could be connected in a straight line.

However, this structure made the placement and wiring process in the CAD flow for generating FPGA configuration data and the bitstream generation process complicated, causing bugs in the CAD flow. In addition, from the layout image of the SLMLET-1 chip shown in Fig. 6.16, it can be seen that there is an unused area at the top of the chip. This indicates that it is possible to effectively utilize the chip's area by making the functional circuits, including the eFPGA-IP, larger in scale.

4.1.4.2 Interface Circuits

When RISC-V, SRAM, and Direct Memory Access Controller (DMAC) communicate with eFPGA, it is necessary to implement an interface circuit on the eFPGA. The same applies when sending and receiving data with an adjacent eFPGA-IP. This can be flexibly designed according to the interface specifications of each RISC-V, SRAM, and DMAC, and even if a bug was discovered, it was easy to debug using the FPGA's circuit reconfiguration function. However, the interface circuit consumes the circuit resources of the eFPGA, which results in a reduced possibility of implementing the application circuit desired by the user. Therefore, to improve this issue, it is necessary to reconsider the implementation method of the interface circuit.

4.1.4.3 Single Operating Clock

The SLMLET-1 chip was operated with a single clock. This meant that the external circuit had to operate at the same frequency as the application circuit implemented in the eFPGA-IP. Therefore, if a circuit with a slow operating frequency is implemented in the eFPGA-IP, the RISC-V processor also operates at the same slow frequency, resulting in the processor's performance not being fully utilized. To improve this issue, it is necessary to review multiple clock domains so that eFPGA and RISC-V can operate in parallel.

4.1.4.4 DSPTILE

When implementing DSPTILE in eFPGA-IP (TEG2), we prioritized short-term design and found it difficult to estimate the circuit area, so we built an 8-bit unsigned multiplier as a dedicated circuit. This made it possible to operate faster than multiplication processing using SLM. However, when processing signed multiplication, it was necessary to convert the negative multiplier and multiplicand to positive numbers once and then multiply them and then to synthesize the logic so that the result would be a negative value in the case of a combination of the signs of the multiplier and multiplicand that would result in a negative value. In addition, an 8-bit multiplier is too small for numerical calculation applications. For DSPTILE, we decided that further high functionality was necessary, as with commercial FPGAs.

4.1.4.5 Logic Synthesis in the CAD Flow

In logic synthesis in the CAD flow of eFPGA-IP (TEG2), it was necessary to use two tools, Yosys and ODIN II, as needed. However, there were cases where the combination of these two tools could not synthesize an equivalent circuit. Therefore, to solve this problem in logic synthesis, it is necessary to reconsider the method.

4.1.4.6 Placing and Routing in the CAD Flow

For the eFPGA-IP (TEG2), an original tool named EasyRouter was developed and used for the routing process. Due to a problem with the EasyRouter specifications, two versions of VPR, 8.1.0 and 8.0.0, were required for the clustering and placement process in the front end of the routing. Furthermore, since the EasyRouter developer has returned to his home country, future maintenance is almost impossible. To solve this problem in placement and routing, the method should be reconsidered.

4.2 The SLMLET-2 and the eFPGA-IP (TEG3)

In this section, we summarize the next version of the SLMLET-2 and the eFPGA-IP (TEG3) in the chip.

4.2.1 The SLMLET-2

The problem with the SLMLET-1 is that, when viewed purely as an SoC chip, both the CPU and FPGA have low performance, making it impossible to implement sufficiently realistic problems [24]. Also, when viewed as a Shell–Role chip, the Shell and Role parts are not connected by a proper standard interface. Taking the above into consideration, the following improvements were made:

- The two-block structure of the eFPGA array was revised to a one-block structure that is much larger than the previous one.

- The RISC-V was replaced with a more full-fledged processor equipped with a cache, the CV32E40S [25]. It was also equipped with a UART to facilitate connection to the host.
- The entire design was IP-based, with IPs connected via OBI (Open Bus Interface) [26, 27].

On the other hand, the following functions confirmed to work on the SLMLET-1 were installed without any changes:

- Direct connection of HyperRAM
- Function to build inter-chip network by HyperBus

As the eFPGA array has grown, the size of configuration data has also increased. Therefore, we decided to set the configuration information by either placing the configuration data on HyperRAM and transferring it or directly transferring it serially from the outside. The hardware migration function of the internal FF data scan chain is also implemented in the SLMLET-2 because it is effective for debugging.

The block configuration of SLMLET-2 is shown on the left of Fig. 6.17. Compared to the SLMLET-1, the SLM has a monolithic large-scale configuration and is connected by a standard OBI switch. The RISC-V cache is a 16-kiB instruction cache and a 16-kiB data cache, which are two-way set associative caches. It is a write-back cache with a block size of 64 B, and data transfer is performed between the HyperRAM and the DMAC. In the SLMLET-2, all modules are connected by standard OBI, but only the cache uses a dedicated interface. It has 16 kiB of built-in shared memory that can be directly accessed from the SLM, which is an FPGA, and

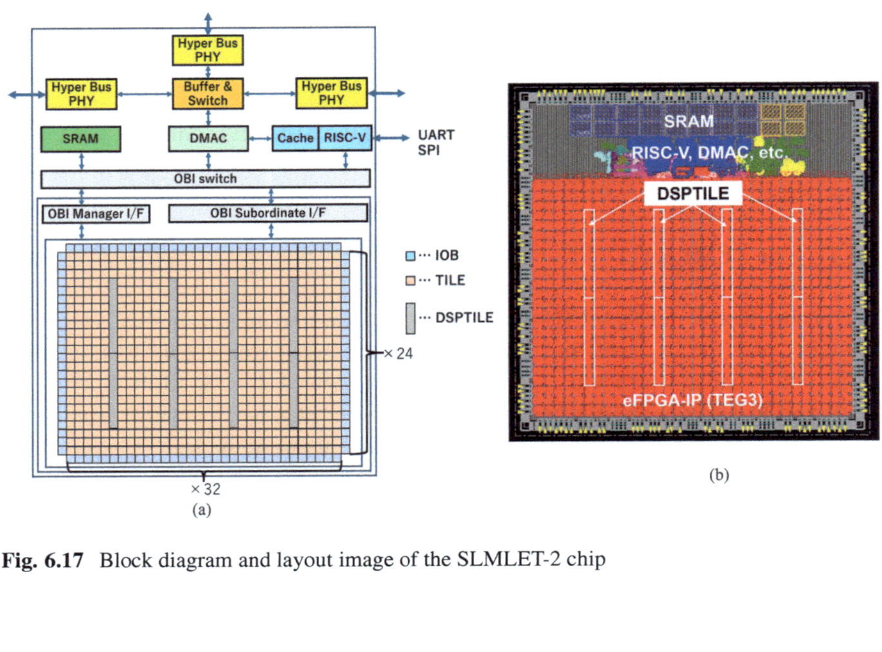

Fig. 6.17 Block diagram and layout image of the SLMLET-2 chip

also has CPU-specific instruction and data memory used for booting and debugging. The SLMLET-2 is implemented using the USJC 55nm LVT process, and it does not have a body bias control function. The chip size is 4.2 mm square, the same as the SLMLET-1, but as shown in the layout image on the right in Fig. 6.17, the eFPGA (TEG3) section occupies approximately 3/4 of the area of the entire chip.

4.2.2 Improvement Policy Toward the eFPGA-IP (TEG3)

In this section, we discuss the improvement policy as follows [28, 29].

4.2.2.1 eFPGA-IP Integration and Scale-Up

To address the issues with the structure of the eFPGA-IP (TEG2) mentioned in the previous section, the array size was changed to 32×24 for one tile array. This change has confirmed that the placement and routing process in the CAD flow and the bitstream generation process, which were previously problematic, have become easier to deal with. Furthermore, comparing the SLMLET-2 layout image shown in Fig. 6.17b with the SLMLET-1 layout image shown in the right of Fig. 6.16, it was confirmed that the unused area of the chip has been effectively utilized.

4.2.2.2 Hard Macro of I/F Circuit

To address the issues in the implementation method of the interface circuit, we decided to create the I/F circuit as a hard macro and place it outside the FPGA part. For the interface, we adopted OBI (Open Bus Interface) from the pulp-platform library provided on GitHub by ETH Zurich and the University of Bologna in Switzerland [27, 27]. With this, the eFPGA-IP (TEG3) operates as a slave via the Subordinate BUS for access from RISC-V and DMAC and as a master via the Manager BUS for SRAM. For this reason, we designed two types of IF circuits: “Subordinate IF” and “Manager IF.”

The Subordinate IF has four read data registers and four write data registers that can be accessed as memory-mapped I/O from RISC-V and DMAC and can send and receive data to and from eFPGA. In addition, after executing the processing requested by RISC-V, the completion of the processing can be notified by polling the status register or by interrupting RISC-V.

The Manager IF has a single transfer function in addition to a burst transfer function so that TEG3 can smoothly send and receive the amount of data requested by SRAM as needed.

When introducing these specifications, it was confirmed during simulation that eFPGA-IP (TEG3) can properly communicate with RISC-V and DMAC via the Subordinate IF and with SRAM via the Manager IF. It was also confirmed that when implementing application circuits on eFPGA-IP (TEG3), the waste of circuit resources is mitigated since the resources consumed by the IF circuit are no longer required.

4.2.2.3 Multiple Clock Domains

To address the issue of a single clock for the entire chip mentioned in the previous section, we designed the eFPGA-IP (TEG3) and its external circuit to operate at two different frequencies so that the performance of the external circuit can be fully utilized. The synchronization of the two clock signals is achieved by inserting a clock domain crossing (CDC) circuit from the pulp-platform library between the eFPGA-IP (TEG3) and its external I/F, so that two clocks of different frequencies can be operated in cooperation. Simulations have confirmed that data communication can be performed normally even when two clocks of different operating frequencies are used for the eFPGA-IP (TEG3) and its external circuit.

4.2.2.4 4-to-1 MUX Function

The issue of reproducing the MUX circuit with the SLMs mentioned in the previous section is addressed by implementing a dedicated circuit for 4-to-1 MUX. This allows us to reduce resource usage and delays compared to using 4-to-1 MUX using only SLM. In the design, the number of LB input signals was expanded from 16 in TEG1 and TEG2 to 20 in TEG3, taking into consideration cases where all input signals of the SLM and 4-to-1 MUX are used. In addition, to accommodate the four 5-SLMs with a total of 20 inputs, four 4-to-1 MUXs were prepared, and the select signal was shared between two pairs, reducing the total number of input signals to 20. This makes it possible to use one TILE as four 4-to-1 MUXs, although there are some restrictions. Furthermore, the output of the BLE can be selected from the 4-to-1 MUX in addition to the SLM and ripple carry adder. Although this improvement slightly increases the area of the TILE, it was possible to layout it within the same size of the SLMLET-1 chip.

4.2.2.5 Multifunctional DSPTILE

The issues with DSPTILE mentioned in the previous section have been addressed with high-performance features in the eFPGA-IP (TEG3) DSPTILE. The DSP block of commercial FPGAs is based on signed multipliers, and by providing redundancy in the bit width, it is possible to support both signed and unsigned numbers, and by simultaneously installing an adder, more advanced calculations are possible [30]. Based on this, the current DSPTILE implements a 17-bit signed multiplier as shown in Fig. 6.18 and can support up to 16-bit multiplication for unsigned numbers. In addition, a 34-bit adder has been added to enable multiplication and accumulation. As a result, the current DSPTILE can select from four main modes: multiply mode, adder mode, multiply and adder mode, and multiply and accumulation mode. The size of the DSPTILE has increased from the previous size of four logic TILEs vertically to the new size of nine logic TILEs vertically. This is because the number of input/output signals to the DSPTILE, including control signals, has almost doubled, and a corresponding number of TILEs are required. Compared to the previous DSPTILE, this configuration has been confirmed to make more effective use of the specified area. This time, eight DSPTILEs are placed in the array configuration of the eFPGA-IP (TEG3), two each in the 6th, 13th, 20th, and 27th columns as shown in Fig. 6.18. This was decided based on the results of a preliminary investigation [31]of the optimal floor plan for DSPTILE, based on the eFPGA-IP (TEG2), which showed that good results were obtained in terms of operating frequency when DSPTILE was used in a pattern where it was concentrated in the center of the array configuration and evenly arranged.

4.2.2.6 Revise of Logic Synthesis Process

Figure 6.19 shows CAD flow for eFPGA-IP (TEG3) that modified from CAD flow for eFPGA-IP (TEG2). To improve the issues with logic synthesis processing described in the previous section, the CAD flow for eFPGA-IP (TEG3) uses only Yosys as a logic synthesizer. This processing can be performed by loading a synthesis library that has been adjusted to be compatible with eFPGA-IP. In addition, Yosys has a variety of subcommands for logic synthesis processing, and by making full use of these, it is possible to generate the type of netlist to use any functional units in eFPGA-IP (TEG3). This makes it possible to reduce resource consumption by configuring the circuit to be configured on the FPGA with area priority, and it is also possible to configure it with speed priority through parallel processing. With the above measures, it has been confirmed that logic synthesis is possible without any problems using only Yosys.

4.2.2.7 Revise of Placement and Routing Process

To alleviate the issues in placement and routing processing described in the previous section, this CAD flow was modified from the eFPGA-IP (TEG2) so that clustering, placement, and routing can be completed using only VPR version 8.1.0. At that time, by adjusting the description of the routing resource graph (rr-graph) file in XML format used by VPR, it is possible to perform routing processing compatible with the original architecture of the eFPGA-IP (TEG3). This will eliminate the need for EasyRouter, which is in a state where future maintenance is almost impossible. VPR also has a wide range of processing subcommands, and by making full use of these, it is possible to adjust how the FPGA configuration circuit is generated. This allows you to configure the circuit to be implemented in the eFPGA-IP with area or speed as the priority. Furthermore, it is also possible to correctly generate the fasm

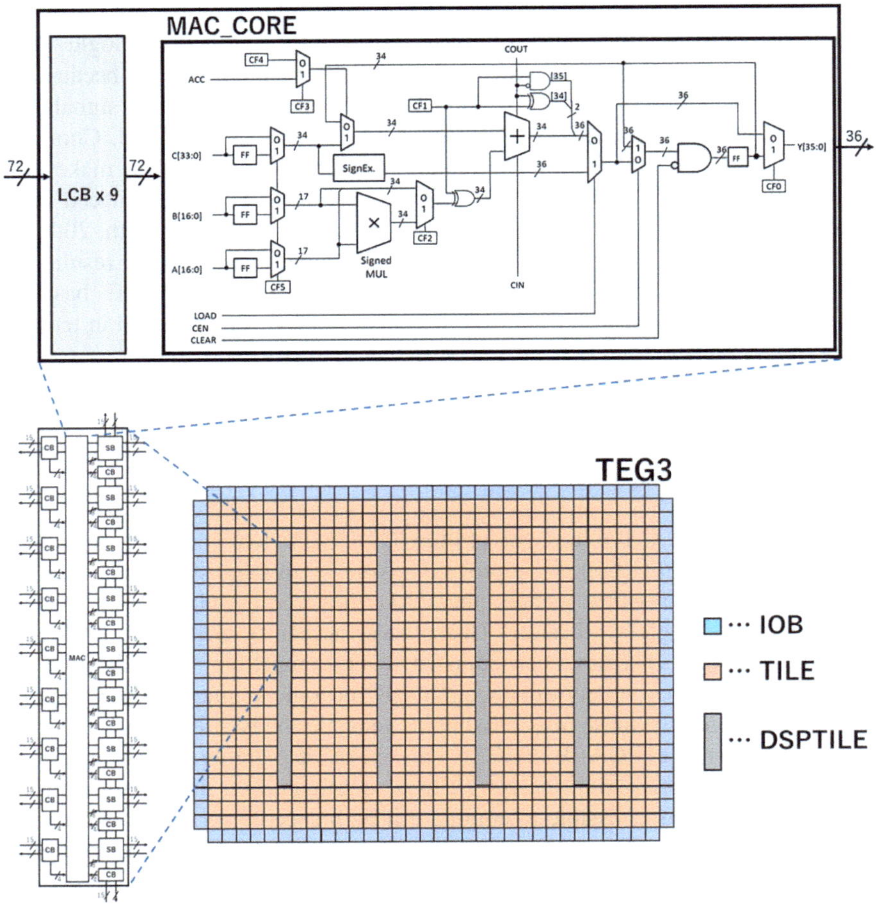

Fig. 6.18 Block diagram and floor plan of DSPTILE on eFPGA-IP (TEG3)

code by appropriately adding meta information for generating a configuration file to the FPGA architecture file and the rr-graph file.

4.2.2.8 Revise of Bitstream Generation Process

As described in the previous section, we have decided not to use EasyRouter in the eFPGA-IP (TEG3) CAD flow, so we need to reconsider the bitstream generation process. The revised CAD flow uses VPR's fasm generator named genfasm, which outputs fasm assembly code. The generated fasm assembly code is parsed using the fasm Python parser provided by the CHIPS Alliance [32], and the back-end part that must be made specifically for the FPGA-IP (TEG3) is newly developed using

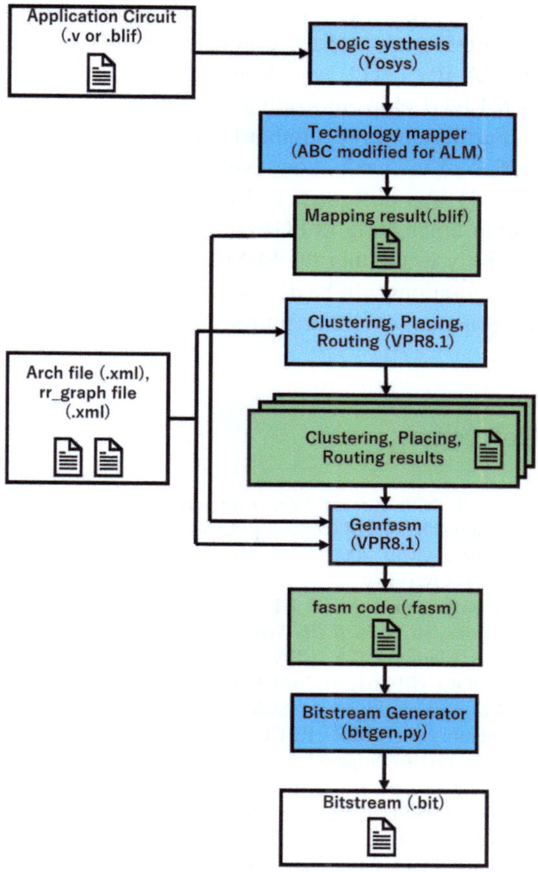

Fig. 6.19 Revised CAD flow for the eFPGA-IP (TEG3)

Python. Using this method, it has been confirmed that a bitstream can be generated without any problems and that the FPGA-IP (TEG3) simulation using this bitstream operates without any problems.

We are currently evaluating the SLMSET-2 chip, which is embedded with eFPGA-IP (TEG3). A future report will describe details of the evaluation results and considerations about the SLMLET-2 chip.

5 Conclusion of This Section

In this section, we first introduced the concept of SLMLET based on a shell-rolled chip for MEC devices. Then, we introduced SLM-based embedded FPGA-IP, which can be implemented with a small number of configuration memories and a small area. Three types of chips were designed and developed as prototypes. The first version, eFPGA-IP (TEG1), was explained in detail. The issues obtained from the consideration of the evaluation results were summarized. Based on the issues, eFPGA-IP (TEG2) was designed, and the second version was implemented as the SLMLET-1 chip embedding eFPGA-IP (TEG2).

Furthermore, we summarized the issues and solutions obtained from the consideration of the evaluation results of SLMLET-1. Based on the solutions, eFPGA-IP (TEG3) was designed, and the third version was implemented as the SLMLET-2 chip embedding eFPGA-IP (TEG3). At the time of writing this report, we are currently evaluating the SLMLET-2 chip. The details of the evaluation results and considerations will be described in a future report. We also plan to proceed with the development environment for the practical use of SLMLET chips.

References

1. ETSI, *Multi-Access Edge Computing* (2018). Accessed 30 Jan 2024. https://www.etsi.org/images/files/ETSITechnologyLeaflets/
2. Xilinx, *Zynq 7000 Data Sheet: Overview*. https://docs.xilinx.com/v/u/en-US/ds190-Zynq-7000-Overview
3. T. Kojima et al., SLMLET: a RISC-V processor SoC with tightly-coupled area-efficient eFPGA blocks, in *2024 IEEE Symposium in Low-Power and High-Speed Chips (COOL CHIPS)*. Los Alamitos, CA, USA (IEEE Computer Society, 2024), pp. 1–6. https://doi.org/10.1109/COOLCHIPS61292.2024.10531161
4. M. Amagasaki et al., SLM: a scalable logic module architecture with less configuration memory (LETTER). IEICE Trans. Fund. Electron. Commun. Comput. Sci. (E99)(A.12), 2500–2506 (2016). https://doi.org/10.1587/transfun.E99.A.2500
5. K. Inoue et al., An easily testable routing architecture and prototype chip. IEICE Trans. Inform. Syst. **E95**(D.2), 303–313 (2012). https://doi.org/10.1587/transinf.E95.D.303
6. M. Kuga et al., An eFPGA generation suite with customizable architecture and IDE. IEICE Trans. Fund. **E106**(A.3), 560–574 (2023). ISSN: 1745-1337. https://doi.org/10.1587/transfun.2022VLP0008
7. J.H. Anderson, Q. Wang., Area-efficient FPGA logic elements: Architecture and synthesis, in *16th Asia and South Pacific Design Automation Conference (ASP-DAC 2011)* (2011), pp. 369–375. https://doi.org/10.1109/ASPDAC.2011.5722215
8. Z. Zilic, Z.G. Vranesic, Using decision diagrams to design ULMs for FPGAs. IEEE Trans. Comput. **47**(9), 971–982 (1998). https://doi.org/10.1109/12.713316
9. M. Iida et al., COGRE: a novel compact logic cell architecture for area minimization. IEICE Trans. Inform. Syst. **E95**(D.2), 294–302 (2012). https://doi.org/10.1587/transinf.E95.D.294
10. K.S. McElvain, IWLS'93 benchmark set: version 4.0, in *Distributed as part of the MCNC International Workshop on Logic Synthesis '93 Benchmark Distribution* (1993)
11. V. Betz, *175.vpr SPEC CPU2000 Benchmark Description File* (2021). http://www.spec.org/osg/cpu2000/CINT2000/175.vpr/docs/175.vpr.html
12. D. Chai, A. Kuehlmann, Building a better Boolean matcher and symmetry detector, in *Proceedings of the Conference Design, Automation and Test in Europe (DATE'06)* (2006), pp. 1079–1084. https://doi.org/10.1109/DATE.2006.243959
13. G.G. Lemieux, D.M. Lewis, Analytical framework for switch block design, in *Proceedings of the Reconfigurable Computing Is Going Mainstream, 12th International Conference on Field-Programmable Logic and Applications*. FPL '02 (Springer, Berlin, 2002), pp. 122–131. ISBN: 3540441085. https://doi.org/10.1007/3-540-46117-5_14

14. E. Ahmed, J. Rose, The effect of LUT and cluster size on deep-submicron fpga performance and density. IEEE Trans. Very Large Scale Integr. Syst. **12**(3), 288–298 (2004). https://doi.org/10.1109/TVLSI.2004.824300
15. M. Amagasaki et al., Physical fault detection and recovery methods for system-LSI loaded FPGA-IP core. IEICE Trans. Inform. Syst. **E100**(D.4), 633–644 (2017). https://doi.org/10.1587/transinf.2016AWI0005
16. J. Rose et al., The VTR project: architecture and CAD for FPGAs from Verilog to routing, in *Proceedings of the ACM/SIGDA International Symposium on Field Programmable Gate Arrays*. FPGA '12. Monterey, California, USA (Association for Computing Machinery, 2012), pp. 77–86. ISBN: 9781450311557. https://doi.org/10.1145/2145694.2145708
17. C. Wolf. *Yosys Open SYnthesis Suite* (2021). http://www.clifford.at/yosys/
18. Berkeley Logic Synthesis and Verification Group, *ABC: A System for Sequential Synthesis and Verification* (2021). http://www.eecs.berkeley.edu/~alanmi/abc/
19. Q. Zhao et al., FPGA design framework combined with commercial VLSI CAD. IEICE Trans. Inform. (E96)(D.8), 1602–1612 (2013). ISSN: 1745-1361. https://doi.org/10.1587/transinf.E96.D.1602
20. OpenSTA, *OpenSTA* (2021). http://opensta.org/
21. M. Kuga, M. Iida, H. Amano, FPL demo: an FPGA-IP prototype chip for MEC devices, in *2022 32nd International Conference on Field-Programmable Logic and Applications (FPL)* (2022), pp. 467–467. https://doi.org/10.1109/FPL57034.2022.00083
22. K. Fujita et al., Advanced channel engineering achieving aggressive reduction of V T variation for ultra-low-power applications, in *2011 International Electron Devices Meeting* (IEEE, 2011), pp. 32–33
23. S. Takagi et al., Applying run-length compression to the configuration data of SLM fine-grained reconfigurable logic. IEICE Trans. Inform. Syst. **E107**(D.12), 1476–1483 (2024). https://doi.org/10.1587/transinf.2024EDP7033
24. H. Amano et al., Shell-role style FPGA chip SLMLET-2 for IoT applications (in Japanese). IEICE Tech. Rep. **124**(188), 60–65 (2024)
25. OpenHW Group, *CV32E40S User Manual* (2024). https://docs.openhwgroup.org/projects/cv32e40s-user-manual/en/latest/
26. Silicon Labs, *OBI 1 ver.1.6.0* (2023), pp. 1–28. https://github.com/openhwgroup/obi/blob/main/OBI-v1.6.0.pdf
27. M. Rogenmoser, T. Benz, *GitHub - pulp-platform/obi* (2024). https://github.com/pulp-platform/obi
28. K. Morihiro et al., Evaluation of the second FPGA-IP prototype for MEC devices and its improvements, in *Proceedings of Asia Pacific Conference on Robot IoT System Development and Platform* (2023), pp. 67–68. https://cir.nii.ac.jp/crid/1050298588047312512
29. Z. Yan et al., Development of improved eFPGA-IP for MEC devices (in Japanese). IEICE Tech. Rep. **124**(188), 66–71 (2024)
30. Xilinx, Inc., *XtremeDSP DSP48A for Spartan-3A DSP FPGAs, UG431 (v1.3)* (2008)
31. K. Takeno et al., Consideration of floorplan of DSP block in FPGA-IP for MEC device (in Japanese). IEICE Tech. Rep. **124**(188), 72–78 (2024)
32. The CHIPS Alliance, *FPGA Assembly (FASM) Parser and Generator* (2024). https://github.com/chipsalliance/fasm

MIX
Papier aus verantwortungsvollen Quellen
Paper from responsible sources
FSC® C105338

If you have any concerns about our products,
you can contact us on
ProductSafety@springernature.com

In case Publisher is established outside the EU,
the EU authorized representative is:
Springer Nature Customer Service Center GmbH
Europaplatz 3, 69115 Heidelberg, Germany

Printed by Libri Plureos GmbH
in Hamburg, Germany